Contents

FOREWORD	iii
INTRODUCTION	1
PART I – BACKGROUND	3
1. What is biodiversity?	3
2. The importance of biodiversity	5
3. Why consider biodiversity in EIAs?	6
4. The UK biodiversity process	7
5. Policy context	9
6. EIA and appropriate assessment	11
7. How is biodiversity different to ecology and nature conservation?	13
8. Current treatment of biodiversity in road EIAs	14
PART II – THE GUIDANCE	16
1. Introduction	16
2. Systematic approach to biodiversity in road EIAs	18
3. Key objective and guiding principles	22
4. Screening and biodiversity	24
5. Scoping biodiversity issues	**28**
5.1 Project activities	31
5.2 Potential impacts	32
5.3 Categories of impacts	35
5.4 Spatial/time issues	36
5.5 Consultees	37
5.6 Scoping outputs	38
6. Baseline conditions	**39**
6.1 Consultees	39
6.2 Relevant information on other projects/activities	39
6.3 New surveys	39
6.3.1 Ecosystem/habitat quality/'biodiversity potential'	40
6.3.2 Key species groups	41
6.3.3 Characteristic species	42
6.3.4 Species susceptible to habitat fragmentation	42
6.3.5 Diversity indices	42
6.4 Biodiversity Information Framework	42
6.5 Evaluation criteria: Assessing the importance of biodiversity elements	49

7. Impact prediction and assessment	**54**
7.1 Methods of impact prediction	54
7.2 Assessment of impact significance	57
7.2.1 NATA based approach	58
7.2.2 Criteria used in recent EISs	62
8. Mitigation and enhancement	**67**
9. Presentation of biodiversity information in EISs	**74**
10. Decision making	**76**
11. Biodiversity monitoring programmes and environmental management plans	**77**

PART III – REVIEW 79

Acknowledgements	**82**
Abbreviations	**83**
Glossary	**84**
Reference Boxes	**92**
General references	**101**
Appendix 1 – Key provisions of existing wildlife policy and legislation	**106**
Appendix 2 – National HAP habitats where road developments are likely to be a factor causing loss or decline	**110**
Appendix 3 – National SAP species where road developments are likely to be a factor causing loss or decline	**112**
Appendix 4 – Cumulative Effects Assessment References	**117**
Appendix 5 – Evaluation Matrix for Determining Impact Significance	**118**
Feedback Form	**119**

Biodiversity Impact

Biodiversity and Environmental Impact Assessment: A Good Practice Guide for Road Schemes

by Helen Byron

August 2000

This Guide is based on research carried out at Imperial College, London, funded jointly by the UK Economic and Social Research Council and the Transport and Biodiversity Group (principally the RSPB, WWF-UK, English Nature, and the Wildlife Trusts). It has been produced by the RSPB with the financial support of WWF-UK, English Nature and the Wildlife Trusts and should be cited as:

Byron, H (2000) *Biodiversity and Environmental Impact Assessment: A Good Practice Guide for Road Schemes*. The RSPB, WWF-UK, English Nature and the Wildlife Trusts, Sandy.

Foreword

Conservation of biodiversity is an essential element of sustainable development and is required by the Convention on Biological Diversity (CBD). In 1994, the UK Government published the *UK Biodiversity Action Plan* and UK actions to conserve biodiversity are continuing. Despite these commitments, we are still losing biodiversity at an unprecedented rate world-wide. Two of the key causes of this loss are habitat loss and fragmentation. Both of these are commonly associated with transport projects, particularly road schemes. In the UK, this problem has been highlighted by several very high profile road schemes, notably Twyford Down and the Newbury Bypass. However, the problem is not restricted to such well known schemes, all road schemes potentially have effects on biodiversity. Concerns that these effects were being over-looked in transport decision-making led our organisations to establish the Transport and Biodiversity Group (TBG).

The TBG identified the potential of Environmental Impact Assessment (EIA) to play an important role in integrating biodiversity considerations into decisions on road schemes, also into development decision-making generally. Indeed, the US Council on Environmental Quality has highlighted the importance of EIA:

'The extent to which biodiversity is considered in future...analyses of federal actions will strongly affect whether biodiversity is adequately protected in the coming decades. It is critical that federal agencies understand and take into account general principles of biodiversity conservation in their decision-making.' (US CEQ, 1993)

The need for guidance on biodiversity in EIA was also strongly recognised at the 18th and 19th annual conferences of the International Association of Impact Assessors (IAIA) held in 1998 and 1999.

This Guide aims to help EIA achieve its potential by providing best practice guidance on the treatment of biodiversity in EIAs for road schemes. It is intended to complement existing guidance and should help all participants in the road EIA process: government, local authorities, planners and ecologists, statutory and nature conservation bodies, developers and promoters, and environmental and ecological consultants involved in the preparation of road Environmental Impact Statements (EISs). The Guide will be particularly relevant to consultants and ecologists planning and carrying out the biodiversity components of EIAs, consultees taking part in the EIA process, and decision-makers evaluating EISs. Although the Guide focuses on road schemes, the principles and detailed guidance it contains can be readily applied to EIAs of other development types.

These principles and advice are strongly grounded in research, being based on work carried out at Imperial College, London. This research involved literature searches, reviews of 40 recent road EISs, and a two stage consultation process with a range of experts in the field of road EIAs. Over 30 experts with a range of perspectives were consulted (government, statutory nature conservation bodies, environmental consultants, non-governmental-organisations, and academics). These experts almost universally thought that there was a strong need for this type of guidance.

We believe that this report can help fulfil the need for such guidance and can play a part in ensuring that potential impacts on biodiversity are thoroughly and systematically assessed in all EIAs. Such assessments will be essential if we are to progress towards our goal of sustainable development. We hope that you find the Guide useful.

Graham Wynne
Chief Executive, RSPB

Robert Napier
Chief Executive, WWF-UK

David Arnold-Forster
Chief Executive, English Nature

Simon Lyster
Director General, The Wildlife Trusts

Introduction

This Good Practice Guide has been developed to improve the consideration of biodiversity in development decision-making by providing best practice guidance on the treatment of biodiversity impacts in Environmental Impact Assessments (EIAs). The Guide provides a detailed approach for road schemes (having been based on an in-depth analysis of recent road EIAs). However, the principles and detailed guidance are applicable for EIAs of all development types.

Otter

> **Objectives of this guidance**
>
> 1. **To provide guidance on a best practice systematic approach for the thorough and consistent assessment of biodiversity in road EIAs.**
> 2. **To provide further best practice guidance on certain weak areas of road ecological impact assessment current practice. Such weak areas include:**
>   - **lack of consideration of the full range of potential impacts,**
>   - **poor baseline surveys/data,**
>   - **lack of explanation of explicit criteria used to determine impact magnitude and significance, and**
>   - **lack of post-project monitoring.**

Box 1

Biodiversity is essential to our lives, providing economic, social and environmental benefits, but, despite this, we appear to be losing biodiversity at an unprecedented rate. In the UK alone over 100 species are thought to have become extinct this century (HM Government, 1994). To conserve and enhance biodiversity it is vital that biodiversity considerations are integrated into all our decision-making. Assessment of biodiversity in EIAs can play an important role in integrating biodiversity into development decisions.

DoE guidance (1995) specifically states '*It is important that a methodical and structured approach is adopted during the EA so that all the potential impacts are covered...*' and this guidance aims to provide that structured approach for the assessment of biodiversity in road EIAs.

Part I of this Guide provides an introduction to biodiversity and an explanation of why it needs to be considered in detail in EIAs. It discusses the concept of biodiversity, how biodiversity differs from the traditional concepts of ecology and nature conservation, the UK biodiversity process, why biodiversity must be considered in EIAs, and current treatment of biodiversity in road EIAs.

Part II provides detailed technical guidance for considering biodiversity in road EIAs. Over-arching principles are explained and advice given on how to deal with biodiversity at each stage of the EIA process. This includes screening criteria for triggering an EIA on biodiversity grounds, scoping checklists to identify potential impacts, assessment of biodiversity baseline conditions, criteria for assessing the magnitude and significance of biodiversity impacts, checklists for identifying potential mitigation and enhancement measures, advice on the presentation of biodiversity information in EISs, and biodiversity monitoring. This guidance will be particularly relevant to consultants and ecologists carrying out EIAs, and decision-makers evaluating the detailed content of EISs.

Part III concludes the guidance by providing a biodiversity checklist. This summaries the good practice treatment of biodiversity in EIA. It is intended for use as a final check to ensure that an EIA has considered all relevant biodiversity issues thoroughly.

Use of the guidance explained in this report should help improve the standard of biodiversity assessment in all road EIAs.

Part I – Background

1. What is biodiversity?

Over the last decade, the buzzword 'biodiversity' has come into widespread use as shorthand for 'biological diversity'. Biodiversity was placed firmly on the international agenda when the Convention on Biological Diversity (CBD) was opened for signature at the 1992 UNEP Earth Summit in Rio Janeiro (UNCED, 1992). 175 countries, including the UK, have now signed the CBD, which came into force on 29 December 1993.

Many definitions of biodiversity have been proposed (see DeLong, 1996; Takacs, 1997), but perhaps the most commonly used is the CBD definition:

> *'The variability among living organisms from all sources including, inter alia, terrestrial, marine and other aquatic ecosystems and the ecological complexes of which they are part; this includes diversity within species, between species and of ecosystems.'*
> (Article 2 CBD, 1992)

For the purposes of this guidance, this definition of biodiversity has been adapted and expanded (based on Noss, 1990) to emphasise the need for EIA to consider all of the levels of biodiversity and the associated structural and functional relationships.

Biodiversity

The total range of variability among systems and organisms at the following levels of organisation:
- bioregional
- landscape
- ecosystem
- habitat
- communities
- species
- populations
- individuals
- genes

and the structural and functional relationships within and between these different levels.

Structural relationships include:	**Functional relationships include:**
connectivity, spatial linkage, patchiness, fragmentation, slope and aspect, the distribution of key physical features (eg outcrops), water availability, dispersion, range and population structure (eg sex and age ratios).	disturbance processes, nutrient cycling rates, energy flow rates, hydrologic processes, human land use trends, demographic processes (eg fertility, survivorship, mortality), metapopulation dynamics, population genetics and population fluctuation.

Box 2

The interconnectedness of all these multiple elements of biodiversity is shown in Figure 1.

Figure 1 – Compositional, structural and functional biodiversity, shown as interconnected spheres, each encompassing multiple levels of organisation

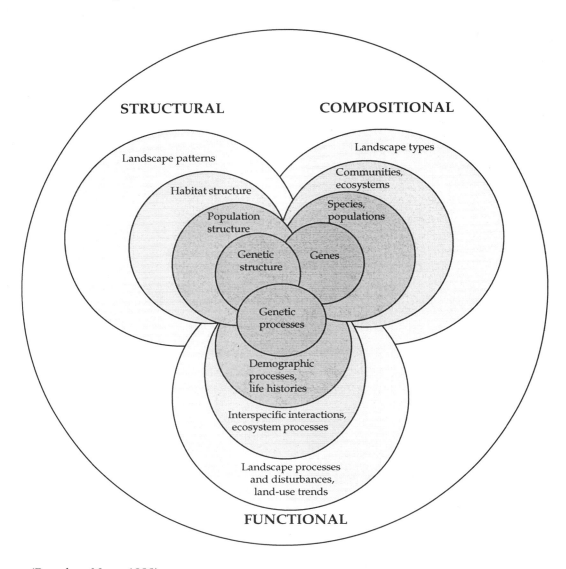

(Based on Noss, 1990).

2. The importance of biodiversity

A great deal has been written about why biodiversity is important[1]. The key reasons why biodiversity matters are summarised in the following box.

Reedbeds

> **Biodiversity matters because:**
>
> - Biodiversity supports life itself.
>
> - Ecosystems can be harvested for economic benefit like food and raw materials.
>
> - Biodiversity can provide indirect economic benefits like flood control or waste water systems.
>
> - Biodiversity has an economic and social value for recreation.
>
> - Biodiversity has aesthetic and spiritual value.
>
> - People value the existence of biodiversity and care whether or not it is conserved.
>
> (RSPB, 1996)

Box 3

---

[1] See, for example, Wilson, 1988; DiSilvestro, 1993; HM Government, 1994; Jeffries, 1997; Reaka-Kudla *et al*, 1997; Takacs, 1997 - in Reference Box 3 and General references.

3. Why consider biodiversity in EIAs?

Biodiversity should be considered in EIAs because conservation of biodiversity is an essential element of sustainable development [2].

'A crucial test of the health of a local environment is whether the wildlife community that is present fully reflects the animal and plant communities normally associated with the habitat in that area. In this way, biodiversity is one of the most important indicators of the state of our environment.' (RTPI, 1999)

It should also be considered because the CBD specifically requires EIAs to consider impacts on biodiversity (Article 14, CBD). Furthermore, considering biodiversity in the development of projects can help ensure their long-term viability.

Article 14 CBD

Article 14 of the CBD, which deals with impact assessment, states:

'Each Contracting Party, as far as possible and as appropriate, shall:

(a) Introduce appropriate procedures requiring environmental impact assessment of its proposed projects that are likely to have significant adverse effects on biological diversity with a view to avoiding or minimising such effects and, where appropriate, allow for public participation in such procedures;

(b) Introduce appropriate arrangements to ensure that the environmental consequences of its programmes and policies that are likely to have significant adverse impacts on biological diversity are duly taken into account; ....'

Box 4

'Biodiversity' is a more holistic and comprehensive approach to considering impacts on 'flora and fauna' and the relationships between them required by the EIA Directive.

---

[2] English Nature, 1993, 1998a; RSPB, 1996; DETR, 1998a.

4. The UK biodiversity process

Implementation of the CBD objectives in the UK should strengthen and broaden the remit of existing wildlife policy and legislation. The primary emphasis of current policy is on the conservation of habitats and species within protected areas. The key provisions of existing policy and legislation are summarised in Appendix 1.

In the UK, as elsewhere around the world, implementation of the obligations of the Convention has focused on the requirement to produce and implement a national biodiversity plan. The key aim of these national plans is to identify priority areas necessary for achieving the CBD's objectives. In 1993, a group of six UK voluntary conservation bodies produced the discussion document *Biodiversity Challenge: an agenda for conservation in the UK* (Wynne *et al*, 1993) (a second edition was published in 1995 (Wynne *et al*, 1995). This document which describes itself as *'a plan for action from the voluntary conservation sector'* was intended to aid the UK Government in the production of a national biodiversity plan. The UK Biodiversity Action Plan (UK BAP) (which drew on the first edition of the Biodiversity Challenge document) was published in January 1994 (HM Government, 1994) with the overall goal:

'To conserve and enhance biological diversity within the UK and to contribute to the conservation of global biodiversity through all appropriate mechanisms'.

UK BAP Objectives for conserving biodiversity

1. To conserve and, where practicable, to enhance:

   a) The overall populations and natural ranges of native species and the quality and range of wildlife habitats and ecosystems.
   b) Internationally important and threatened species, habitats and ecosystems.
   c) Species, habitats and managed ecosystems that are characteristic of local areas.
   d) The biodiversity of natural and semi-natural habitats where this has been diminished over recent past decades.

2. To increase public awareness of, and involvement in, conserving biodiversity.
3. To contribute to the conservation of biodiversity on a European and global scale.

(HM Government, 1994)

Box 5

The UK Biodiversity Group identified a list of Species of Conservation Concern which fall in one or more of the following categories:

- Threatened endemic and other globally threatened species;
- Species where the UK has more than 25% of the world or appropriate biogeographical population;
- Species where numbers or range have declined by more than 25% in the last 25 years;
- In some instances where the species is found in fewer than 15 ten km squares in the UK; and
- Species which are listed in the EU Birds or Habitats Directives, the Berne, Bonn or CITES Conventions, or under the Wildlife and Countryside Act 1981 (WCA 1981) and the Wildlife Order (Northern Ireland) 1985 (English Nature, 1998b- Reference Box 3).

Certain species within the list of Species of Conservation Concern have been classified as Priority Species. These are species which are globally threatened and/or species which are rapidly declining in the UK, ie by more than 50% in the last 25 years (English Nature, 1998b - Reference Box 3).

Costed action plans (Habitat Action Plans (HAPs) and Species Action Plans (SAPs)) or conservation statements (Habitat Statements (HSs) and Species Statements (SS)) have now been produced for all the Priority Species (over 450 species) and 35 habitat types[3]. The targets and proposals put forward in the plans are designed to be appropriate up to 2010.

As well as these national level initiatives, the UK BAP is also being implemented through a series of local biodiversity action plans (LBAPs) for priority habitats and species. English Nature has developed a Natural Area's approach to guide its conservation work and to provide a framework for the LBAP process[4]. Various LBAPs, both at regional and local authority levels, are at different stages of preparation eg Hampshire, Kent, the South West and Mendip District have published Action Plans and Bradford City Metropolitan Council and Surrey are in the process of preparing plans.[5] These national and local action plans are key references and sources of information for any EIA.

The Scottish Executive has recently issued a consultation version of a *Trunk Road Biodiversity Action Plan (TRBAP)* (Scottish Executive, 1999a). This document states that it has two purposes:

'To assist in the delivery of biodiversity targets and objectives as set down in the Scottish Local Biodiversity Action Plans' and *'To raise awareness of biodiversity in all engineers, managers, planners, designers and ecologists working on the Scottish trunk road network so that protecting our natural heritage can become part and parcel of everyday work'.*

While the idea of a TRBAP is supported, it is hoped that the final version of the document will be much stronger than the consultation version to enable it to go further towards fulfilling these purposes. It is understood that the Highways Agency is also planning to prepare a TRBAP and that work on this is at a preliminary stage.

---

[3] HM Government, 1995b; English Nature, 1998b and c, 1999a, b, c and d - see Reference Box 3.
[4] There are 120 Natural Areas each with a unique identity on the basis of the wildlife and natural features of the landscape and the opinions of local people (English Nature, 1997a, 1998d). Detailed profiles have been produced for each Natural Area. The aim is that the Natural Areas will help the breakdown of national HAP and SAP targets to a more local level (English Nature, 1998d). As part of this process, Natural Area reports have been produced for each of the English regions (English Nature, 1999g and h) - see Reference Box 3.
[5] A database of LBAPs and relevant contacts is available from the UK Biodiversity Secretariat (DETR, 1999a) and is also on the Secretariat's website at http://www.jncc.gov.uk/ukbg.

5. Policy context

Although production of the UK BAP and supporting action plans is at an advanced stage, there is no obvious integration of biodiversity obligations in key sectoral policies. For example there is a lack of guidance for incorporating biodiversity considerations into EIA, Strategic Environmental Assessment, policy appraisal, etc. (as mentioned in Article 14 CBD).

Current Government guidance on nature conservation and planning[6] does not explicitly address interfaces with the UK biodiversity process. Indeed, only the planning guidance on Natural Heritage in Scotland (NPPG 14) explicitly discusses the UK and LBAPs noting that: *'planning authorities can make an important contribution to the achievement of biodiversity targets by adopting policies which promote and afford protection to species and habitats identified as priorities in LBAPs'* (SOED, 1999). There is a need for further Government guidance on this issue, in particular on what weight the UK and LBAPs are to be given in the planning system. Such guidance could be incorporated in revised versions of the existing planning guidance on nature conservation, such as the planned revisions of PPG 9 and TAN 5.

Biodiversity is not explicitly mentioned in UK EIA legislation. This may be largely explained by the historical timing of EIA and the CBD. The EC EIA Directive was agreed in 1985 before the CBD. Neither the EIA Directive nor the EIA Amendment Directive (which was agreed in 1997) explicitly mention biodiversity. However, the preamble to the EIA Directive does refer to the need to assess *'effects of a project on the environment...to ensure maintenance of the diversity of species and to maintain the reproductive capacity of the ecosystem as a basic resource for life'* (CEC, 1985). Further, as the EIA Directive requires the identification, description and assessment of direct and indirect effects of a project on flora and fauna and the interaction between these and soil, water, air, climate and the landscape, taking a purposive approach to the legislation, it is clear that the treatment of biodiversity is an integral part of EIA.

Great crested newt

EIA literature reveals that some components of biodiversity – specifically endangered species and habitat loss – are addressed in most EIA studies where they are relevant, but that EIAs are less likely to address other aspects of biodiversity such as diversity at the genetic and ecosystem levels, diversity of non-threatened species, diversity within species, and the functional components of biodiversity[7]. So it appears that components of biodiversity which are already protected (protected areas or status) are more likely to be included in EIA than components which hold less popular status but may be important to the long-term productivity of ecosystems and maintenance of biodiversity (Bagri *et al*, 1998).

Several commentators have acknowledged the need to amend existing EIA practice to encompass the full range of biodiversity impacts[8]. The RSPB believe that EIA is important for biodiversity in the UK context: *'The UK Biodiversity Action Plan emphasises the Government's intention to take account of sustainability and biodiversity conservation objectives in the land-use planning system. We see EA as a key tool through which to achieve this intention.'* (RSPB, 1995).

Some guidance on biodiversity in EIA has been issued outside the UK (see Reference Box 1). However, at present there is no UK guidance to help this process. More generally in the UK, a guide to biodiversity for the planning and development sectors in the South West has been published (ALGE *et al*, 2000) and the Royal Town Planning

---

[6] PPG 9 (DoE, 1994), NPPG 14 (SOED, 1999) and Circular 6/1995 (SOED, 1995), TAN 5 (Welsh Office, 1996), and PPS 2 (DoE-Northern Ireland, 1997).
[7] Bagri *et al*, 1998; Le Maitre *et al*, 1997; Sadler, 1996; Hirsch, 1993.
[8] Hirsch, 1993; UNEP, 1998a and b; Bagri *et al*, 1998; IAIA, 1998; IAIA 1999 Biodiversity Working Group, unpublished.

Institute (RTPI) has published *Planning for Biodiversity: Good Practice Guide* (RTPI, 1999). The RTPI's guide states:

'The effects on biodiversity should be assessed in every statutory environmental statement and considered throughout the environmental assessment process, particularly at the scoping stage. Even where there may be no significant adverse effects on biodiversity the environmental assessment process may highlight opportunities for enhancement. In some cases it may be considered that these help to offset some adverse effects unrelated to wildlife.'
(RTPI, 1999)

6. EIA and appropriate assessment

In relation to Natura 2000 sites (Special Areas of Conservation (SACs) and Special Protection Areas (SPAs) and proposed or candidate sites), Article 6(3) of the Habitats Directive requires that an appropriate assessment of any plans or projects on a site's conservation objectives must be carried out to ensure that the integrity of the site is not adversely affected. (See Box 6 below and also section 7.2.1 which discusses the concept of 'integrity'). This will obviously entail evaluating the impacts of the proposal on the site itself, but may also require a consideration of the impacts on the feature(s) of interest in a wider context such as the regional or national level.

Article 6 Habitats Directive

(For a summary of the Habitats Directive generally see Appendix 1)

Articles 6(3) and 6(4) set out the circumstances in which plans and projects with negative effects may or may not be allowed.

'Any plan or project not directly connected with or necessary to the management of the site but likely to have a significant effect thereon, either individually or in combination with other plans or projects, shall be subject to appropriate assessment of its implications for the site in view of the site's conservation objectives. In light of the conclusions of the assessment of the implications for the site and subject to the provisions of paragraph 4, [Article 6(4) below], the competent national authorities shall agree to the plan or project only after having ascertained that it will not adversely affect the integrity of the site concerned and, if appropriate, after having obtained the opinion of the general public.' Article 6(3)

'If, in spite of a negative assessment of the implications for the site and in the absence of alternative solutions, a plan or project must nevertheless be carried out for imperative reasons of overriding public interest, including those of a social or economic nature, the Member State shall take all compensatory measures necessary to ensure that the overall coherence of Natura 2000 is protected. It shall inform the Commission of the compensatory measures adopted.

Where the site concerned hosts a priority natural habitat type and/or a priority species, the only considerations which may be raised are those relating to human health or public safety, to beneficial consequences of primary importance for the environment or, further to an opinion from the Commission, to other imperative reasons of overriding public interest.' Article 6(4)

The European Commission has recently published an interpretation guide on the provisions of Article 6, which aims to ensure that the provisions are applied consistently throughout the European Community. This guide includes helpful discussion of:

- What is meant by *'plan or project not directly connected with or necessary to the management of the site'*;
- How to determine whether a plan or project is *'likely to have a significant effect'*;
- What is meant by *'appropriate assessment of its implications for the site in view of the site's conservation objectives'*;
- The adoption of *'compensatory measures'*; and
- What happens with sites hosting priority habitats and/or species.

(European Commission, 2000)

Box 6

Where such a plan/project does not require an EIA pursuant to UK EIA legislation the appropriate assessment will be carried out as a 'stand alone' exercise. This will generally be more focused than an EIA in that it will specifically consider the implications of the plan/project for the site's conservation objectives. However, there will be cases where a project will require both an EIA under the EIA legislation and an appropriate assessment under the Habitats Directive. In these circumstances, the EIA for the project could incorporate the appropriate assessment. Such an approach was adopted in the EIA for a proposed Welsh road scheme (A465 Abergavenny to Hirwaun Dualling) where the EIS (Welsh Office Highways Directorate, 1997) included the appropriate assessment in the form of a specific section looking at the potential effects of the project on the integrity of a candidate SAC (cSAC). Where this approach is adopted, to avoid any possible confusion, the relevant section of the EIS should be clearly identified as comprising the appropriate assessment.

7. How is biodiversity different to ecology and nature conservation?

'Biodiversity' is used both as a broad political term (as shorthand for the living life support systems of the world) and in a more scientific and technical sense eg as defined by Noss (1990) and reflected in the definition adopted for the purposes of this guidance.

The term biodiversity is being used as a wider concept to provide fresh impetus for nature conservation in the form of a new framework and funding. Importantly, it also includes the concept of sustainable use as a core component of, and tool for, the conservation of biodiversity.

The more scientific approach emphasises the need to understand the different levels of biological units, the different scales they operate at, the links they provide and the functions they fulfil. In this sense, it refocuses the concepts of ecology and nature conservation away from a traditional species based approach, towards a more holistic approach which explicitly considers whole ecosystems and landscape/bioregional scales (Takacs, 1997).

8. Current treatment of biodiversity in road EIAs

Impacts on biodiversity are not currently considered explicitly in road EIAs (Byron & Sheate, 2000; Byron *et al*, 2000). For example, none of the 40 recent road EISs reviewed by Byron *et al* (2000) specifically referred to potential impacts on biodiversity. The current weaknesses in road EIAs in relation to biodiversity (such as those summarised in Box 7) may mean that major effects on biodiversity are missed.

Biodiversity and road EIAs

Current weaknesses include the lack of:
- Use of biodiversity terminology/linkages with UK BAP, HAPs, HSs, SAPs and LBAPs.
- Proper consideration of non-designated sites.
- Consideration of non-protected species.
- Consideration of all levels of biodiversity eg focus on site scale rather than ecosystem level.
- Consideration of structural/functional relationships.

(Byron *et al*, 2000)

Box 7

Bumblebee

Current UK guidance on road EIAs[9] and guidance on EIA generally[10] largely predates the UK biodiversity process and has not been updated to refer explicitly to impacts on biodiversity and the potential interfaces with the UK biodiversity process. It is however understood that, the main UK guidance on road EIAs (DoT, 1993) is due to be revised this year and that these revisions will incorporate biodiversity issues.

This lack of guidance could be a major factor in the current poor treatment of biodiversity issues in road EIAs. Le Maitre *et al* (1997) reported that *'Many interested and affected parties [of EIA], and often the personnel leading environmental impact assessments, do not understand the full meaning of biodiversity'* and that this situation is often exacerbated by those involved being given inadequate terms of reference for addressing biodiversity issues. To improve this situation, Bagri *et al* (1998) recommend that guidelines for incorporating biodiversity into EIA in practice are developed. Research has shown that EIA guidance can have a positive effect on the quality of EISs (CEC, 1996; Donnelly *et al*, 1998; Geraghty, 1999). Geraghty (1999) notes that for EIA guidance to be used, the presentation of the guidance is as important as the nature of the guidance.

The consideration of biodiversity in EIA is in its infancy, whereas ecological impact assessment has been a fundamental part of EIA for many years. However, despite existing guidance on the treatment of ecological impacts in EIAs (see Reference Box 2), there are some aspects where current practice is poor (Treweek *et al*, 1993; Byron *et al*, 2000) and where further guidance would be helpful - see Box 8.

---

[9] Box & Forbes, 1992; Department of Transport (DoT), 1993; English Nature, 1994a - see Reference Box 2.
[10] English Nature, 1994b; DoE, 1989, 1995; RSPB, 1995; Morris & Therivel, 1995 - see Reference Box 2.

> **Ecology and road EIAs: current weaknesses**
>
> - Lack of consideration of full range of impacts, especially indirect and cumulative impacts.
> - Lack of explanation of the criteria used to determine impact magnitude.
> - Lack of explanation of the criteria used to determine impact significance.
> - Lack of consideration of the full range of possible mitigation measures.
> - Lack of consideration of possibilities for enhancement.
> - Poor baseline surveys/data.
> - Poor interpretation of results.
> - Poor use of relevant scientific literature.
> - Poor presentation of information in EISs.
> - Lack of post-project monitoring.
>
> (Treweek *et al*, 1993; Byron *et al*, 2000)

Box 8

Part II – The Guidance

1. Introduction

This guidance is intended to improve the consideration of biodiversity in road EIAs. It combines elements of current best practice ecological impact assessment with guidance on a systematic approach for considering biodiversity issues within road EIAs. The context for the assessment is provided by a key objective and guiding principles for biodiversity. The guidance considers biodiversity issues relevant to each stage of the EIA process including:

- **Screening** – What biodiversity considerations should trigger a road EIA?

- **Scoping** – What alternatives should be considered? What activities may lead to impacts on biodiversity? What elements of biodiversity might be affected?

- **Baseline conditions** – Useful sources of background information on biodiversity. What new surveys should be carried out? What criteria should be used to evaluate the relative importance of different biodiversity elements?

- **Impact prediction and assessment** – What impact prediction techniques are appropriate for biodiversity impacts? What criteria should be used to assess the magnitude of biodiversity impacts? What criteria should be used to evaluate the significance of biodiversity impacts?

- **Mitigation and enhancement** – What mitigation/enhancement measures should be considered?

- **EIS preparation** – How should the biodiversity information be presented?

- **Decision-making** – Consideration of the biodiversity information presented in the EIS.

- **Biodiversity monitoring and environmental management plans** – What elements of biodiversity should be monitored? What information sources could this monitoring information be fed into? Should an environmental management plan be established?

Use of the systematic guidance set out in the following sections should ensure that biodiversity issues are given improved consideration in road EIAs/EISs. Some of the differences from current EIAs/EISs are highlighted in Box 9.

How will an EIA carried out following this guidance be different from current EIAs?

- The EIA will look at all the relevant levels of biodiversity ie bioregional, landscape, ecosystem, habitat, communities, species, populations and where appropriate individuals and genes. (IAIA 1999 Biodiversity Working Group (unpublished) stressed the need for EIA to move away from its current over emphasis on habitats and species in protected areas).

- The EIA will consider connections between the levels of biodiversity ie will look at structural and functional relationships (such as connectivity, fragmentation and disturbance, hydrologic and demographic processes). Currently there is often little emphasis on processes (other than hydrology) within EIAs.

- The EIA will use biodiversity terminology and will explicitly tie in with the UK BAP process, using the BAP targets (at both national and local levels) to provide context for the assessment and to help set appropriate criteria for judging impact magnitude and significance.

- The systematic approach may require more information to be collected on certain aspects of biodiversity than for many current EIAs, but the emphasis is **not** on surveying everything in detail. Instead, the systematic approach provides the structure for focusing on the biodiversity receptors that are important and should be studied in more detail. It enables the level of information needed to form the basis of impact predictions to be collected. In many current EIAs, the level of information collected (eg species lists) is not sufficient to make impact predictions.

- The EIA will consider the full range of potential impacts including indirect, cumulative and induced impacts. It will not focus solely on direct losses of habitats and species as is current practice in many road EIAs.

- The EIS will be as clear as possible about the predicted impacts ie these will be quantified wherever possible, timescales will be indicated and confidence/uncertainty in predictions will be stated.

- The EIA will set out explicit criteria used for judging impact magnitude and significance. It will explain how these have been derived and how they correlate with relevant targets eg BAP targets.

- The EIS will be clear about proposed avoidance, mitigation, compensation and enhancement measures. It will not use confusing terminology that appears to try and disguise compensation/mitigation measures by describing them as enhancement. This will enable the EIA to give a clear indication of the overall effect of the scheme. The proposed likelihood of success of mitigation, compensation and enhancement measures and the timescales involved will be discussed.

- The EIS will not be presented as the end of the EIA process. The EIA will include procedures eg environmental management plans (to provide a framework for the on-going management of the road) and monitoring, to enable evaluation the EIS predictions and facilitate adaptation of management regimes/mitigation measures as necessary.

Box 9

2. Systematic approach to biodiversity in road EIAs

'The steps required to evaluate effects on biodiversity are basically those of traditional highway impact assessment applied with a landscape perspective and specific biodiversity endpoints' (Southerland, 1995).

This guidance proposes a systematic framework approach to the treatment of biodiversity in road EIAs. The biodiversity issues that should be considered at each stage in the EIA process are shown in Figure 2. A key element of this approach is that examination of biodiversity issues in road EIAs should take place in the context of the key objective and guiding principles for biodiversity. These are explained further below. Adoption of this systematic approach will ensure that biodiversity considerations are thoroughly treated at each stage of a road EIA.

Different EIA Regulations apply to UK roads that are planned by central Government and those that are planned by local government or privately and depending on location within the UK. Table 1 lists the various Regulations that apply. Table 2 summarises the sections of the various Regulations that apply at each of the stages of the EIA process shown in Figure 2. It also indicates the sections of this guidance where each of these stages is discussed in detail.

Figure 2 – Systematic approach to biodiversity issues in road EIA

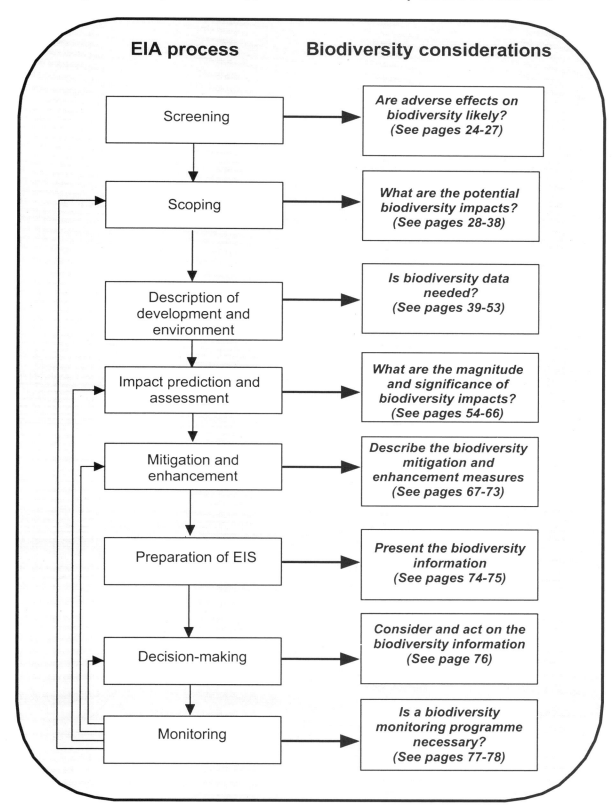

Table 1 – UK EIA Regulations

Regulations	The Highways (Assessment of Environmental Effects) Regulations 1999 (SI 1999 No. 369)	The Town and Country Planning (Environmental Assessment) (England and Wales) Regulations 1999 (SI 1999 No. 293)	The Environmental Impact Assessment (Scotland) Regulations 1999 (Scottish SI 1999 No. 1)	Roads (Environmental Impact Assessment) Regulations (Northern Ireland) 1999 (SR 1999 No. 89)	The Planning (Environmental Impact Assessment) Regulations (Northern Ireland) 1999 (SR 1999 No. 73)
Where applicable	England and Wales	England and Wales	Scotland	Northern Ireland	Northern Ireland
The type of road to which applicable	Centrally planned (ie motorways and trunk roads)	Local government/ privately planned roads	Part II of the Regs. applies to local government/ privately planned roads

Part III of the Regs. applies to centrally planned roads | Centrally planned | Local government/ privately planned roads |
| Competent authority (ie the authority which determines an application for a project to proceed | The Secretary of State (SoS) for the Environment, Transport and the Regions for projects in England and the SoS for Wales for projects in Wales | The relevant local planning authority | The relevant local planning authority for local government/ privately planned roads

The Scottish Minister for centrally planned roads | Northern Ireland Department of the Environment | Northern Ireland Department of the Environment |
| Relevant Government guidance on application of the Regulations | | Circular 02/99 dated 12 March 1999 (DETR, 1999b) | Circular 15/1999 dated August 1999 contains guidance on Part II of the Regs (Scottish Executive, 1999b)

Planning Advice Note (PAN) 58 – Environmental Impact Assessment (Scottish Executive, 1999c) | | Development Control Advice Note 10 – Environmental Assessment (DoE-Northern Ireland, 1989)

Planning Policy Statement (PPS) 3 Development Control: Roads Considerations (DoE-Northern Ireland, 1996) |

Table 2 – Sections of UK EIA Regulations

Stage of EIA process and relevant sections of this guidance	The Highways (Assessment of Environmental Effects) Regulations 1999 (SI 1999 No. 369)	The Town and Country Planning (Environmental Impact Assessment) (England and Wales) Regulations 1999 (SI 1999 No. 293)	The Environmental Impact Assessment (Scotland) Regulations 1999 (Scottish SI 1999 No. 1)	Roads (Environmental Impact Assessment) Regulations (Northern Ireland) 1999 (SR 1999 No. 89)	The Planning (Environmental Impact Assessment) Regulations (Northern Ireland) 1999 (SR 1999 No. 73)
Screening Section 4 – pages 24-27	Sections 105A(2), 105B(1) & 105B(2)	Regulations 4, 5 & 6	Part II Regulations 4, 5 & 6 Part III Sections 20A(2), 20A(3) & 20A(4)	Sections 67(3), 67A(1) & 67A(2)	Regulations 3, 5 & 6
Scoping Section 5 – pages 28-38	Sections 105A (4) & 105A(5)	Regulations 10 & 11	Part II Regulations 10 & 11 Part III Sections 20A(7) & 20A(8)	Sections 67(5) & 67A(6)	Regulation 6
EIS preparation Sections, 6, 7, 8, 9 & 11 – pages 39-75 & 77-78	Sections 105A(4) & 105A(5) & Annex IV of the Amended EIA Directive	Regulation 12	Part II Regulation 12 Part III Sections 20A(7) & 20A(8) & Annex IV of the Amended EIA Directive	Sections 67(5) & 67(6) & Annex IV of the Amended EIA Directive	Regulation 7
EIS submission and publication Section 9 – pages 74-75	Section 105A(3)	Regulations 13- 19	Part II Regulations 13 –19 Part III Section 20A(2)	Section 67(4)	Regulations 9-16
Consultation and participation Sections 5, 6 & 11 – pages 28-53 & 77-78	Sections 105B(3) & 105B(4)	Regulation 12	Part II Regulation 12 Part III Section 20A(5) & 20A(6)	Sections 67A(3), 67A(4), 67A(10), 67A(5) & 67A(6)	Regulation 14
Decision making Section 10 – page 76	Sections 105B(5), 105B(6) & 105B(7)	Regulation 3	Part II Regulation 3 Part III Sections 20A(5) & 20A(6)	Sections 67A(7), 67A(8) & 67A(9)	Regulation 4

Note: *The provisions of the Highways (Assessment of Environmental Effects) Regulations 1999, Part III of The Environmental Impact Assessment (Scotland) Regulations 1999, and the Roads (Environmental Impact Assessment) Regulations (Northern Ireland) 1999 are referred to as 'sections' because these substitute new sections into the Highways Act 1980, The Roads (Scotland) Act 1984, and Part V of the Roads (Northern Ireland) Order 1993 respectively. As the Town and Country Planning (Environmental Assessment) (England and Wales) Regulations 1999, Part II of The Environmental Impact Assessment (Scotland) Regulations 1999, and The Planning (Environmental Impact Assessment) Regulations (Northern Ireland) 1999 are 'stand-alone' they are referred to as Regulations.*

3. Key objective and guiding principles

Most road projects will inevitably lead to some loss of biodiversity but this can be minimised by full use of impact avoidance, mitigation and compensation measures. Furthermore, road projects potentially offer opportunities to enhance biodiversity and contribute to the achievement of HAP/SAP targets. Road EIAs should adopt the positive approach to biodiversity outlined in the key objective. 'Significance' is considered in detail in section 7.

> **Key objective**
>
> To ensure that road schemes:
>
> 1. Do not significantly reduce biodiversity at any of its levels; and
> 2. Enhance biodiversity wherever possible.

Box 10

The principles that should guide the consideration of biodiversity in road EIAs are set out below. These principles can act as 'assessment end points' for road EIAs. Ie the final EIS can be compared to these principles to evaluate whether or not the EIA process has fully considered biodiversity issues and resulted in a scheme that will not significantly reduce biodiversity and which will incorporate biodiversity enhancements wherever possible. This is discussed in more detail in the Part III. Individual principles are explained further in later sections of this guidance. However, the principles are summarised here to provide a checklist of good biodiversity practice for road EIAs that can be referred to throughout the EIA process.

The key objective and guiding principles are built upon principles from the CBD, the UK BAP, existing guidance on biodiversity in impact assessment[11] and comments received during the two-stage consultation process carried out as part of the development of this guidance.

Bittern

---

[11] US CEQ, 1993; CEAA, 1996a; World Bank, 1997 - See Reference Box 1.

Guiding Principles

✓ Avoid impacts on biodiversity and create opportunities for enhancement of biodiversity wherever possible by route selection and scheme design. Where this is not possible identify the best practical mitigation and enhancement option to ensure that there is no significant loss of biodiversity. Compensation measures such as translocation should be viewed as a last resort.

✓ Apply the precautionary principle to avoid irreversible losses of biodiversity. ie where an activity raises threats or harm to biodiversity precautionary measures should be taken even if certain cause and effect relationships are not scientifically established.

✓ Widen existing EIA practice to an ecosystem perspective - ie consider the impacts of a road scheme on biodiversity **and** possible enhancements of biodiversity in the context of local and regional ecosystems, not just the immediate vicinity of the road.

✓ Safeguard genetic resources by protecting the higher levels of biodiversity (ie individuals, populations, species, and communities, etc.) and the environmental processes which sustain them.

✓ Consider the full range of impacts on biodiversity eg indirect and cumulative impacts not just the direct impacts such as species and habitat loss.

✓ The study area of the scheme should reflect the impact type (eg indirect effects will often extend throughout a watershed) rather than taking a fixed width corridor approach.

✓ Evaluate the impacts of a road scheme on biodiversity in local, regional, national, and, where relevant, international contexts ie an impact could be minor locally but significant at a national level eg where the locality has a very high proportion of a nationally rare biodiversity resource.

✓ Retain the existing pattern and connectivity of habitats eg protect natural corridors and migration routes and avoid artificial barriers. Where existing habitat is fragmented implement measures eg tunnels, bridges to enhance connectivity.

✓ Use buffers to protect important biodiversity areas wherever possible.

✓ Maintain natural ecosystem processes in particular hydrology and water quality. Wherever possible use soft engineering solutions to minimise impacts on hydrology.

✓ Strive to maintain/enhance natural structural and functional diversity eg ensure that the quality of habitats and communities is not diminished and wherever possible is enhanced by the road scheme.

✓ Maintain/enhance rare and ecologically important species (key species) - ie protected species, SAP species, characteristic species for each habitat as loss of these may affect a large number of other species and can affect overall ecosystem structure and function.

✓ Decisions on biodiversity should be based on full information and monitoring must be part of the EIA process. The results of monitoring should be available to allow evaluation of the accuracy of impact prediction and should be widely circulated to help improve future road scheme design and mitigation.

✓ Implement on-going management plans for existing and newly created habitats and other mitigation, compensation and enhancement measures.

Box 11

4. Screening and biodiversity

Screening is the process of determining whether or not an EIA is necessary.

When is an EIA required for a road project?

UK motorways and trunk roads - The Highways (Assessment of Environmental Effects) Regulations 1999, Part III of The Environmental Impact Assessment (Scotland) Regulations 1999, or the Roads (Environmental Impact Assessment) Regulations (Northern Ireland) will apply depending on the location of the project (see Table 1). An EIA is required for:

1. Construction of motorways and express roads.
2. Construction of a new road of 4 or more lanes, or re-alignment and/or widening of an existing road so as to provide 4 or more lanes, where such new road or re-alignment and/or widened section of road would be 10 km or more in a continuous length.
3. Where:
   - any part of the road development will be carried out in a sensitive area (sensitive areas include internationally and nationally designated nature conservation sites)

   OR
   - the area of the proposed works (which 'includes any area occupied by apparatus, equipment, machinery, materials, plant, spoil heaps or other facilities or stores required for construction or installation') exceeds 1 hectare

   AND
   - the road project is likely to give rise to **'significant environmental effects'**

   OR
   - the road project is the construction or improvement of a special road.

For each particular project a determination as to whether or not an EIA is required must be published by the Secretary of State or (in Northern Ireland) the Department of the Environment.

Local authority and private roads - Town and Country Planning (Environmental Impact Assessment) (England and Wales) Regulations 1999, Part II of The Environmental Impact Assessment (Scotland) Regulations 1999, or The Planning (Environmental Impact Assessment) Regulations (Northern Ireland) 1999 will apply depending on the location of the project (see Table 1). An EIA is required if:
- any part of the road development will be carried out in a sensitive area (sensitive areas include internationally and nationally designated nature conservation sites)

OR
- the area of the proposed works exceeds 1 hectare

AND
- the road project is likely to give rise to **'significant environmental effects'**.

If the developer is uncertain if a road project falls within these criteria it can decide to carry out an EIA anyway, or ask the relevant planning authority for a screening opinion as to whether an EIA is required.

Box 12

> **What are 'significant effects on the environment'?**
>
> The EIA Directive (CEC, 1985) as amended by the EIA Amendment Directive (CEC, 1997) sets out in Schedule III the criteria to be considered when determining whether an EIA is needed:
>
> **1. Characteristics of projects**
> The characteristics of projects must be considered having regard, in particular, to: the size of the development, the cumulation with other projects, the use of natural resources, the production of waste, pollution and nuisances, and risk of accidents, having regard in particular to substances or technologies used.
>
> **2. Location of projects**
> The environmental sensitivity of geographical areas likely to be affected by projects must be considered, having regard, in particular, to: the existing landuse, the relative abundance, quality and regenerative capacity of natural resources in the area, the absorption capacity of the natural environment paying particular attention to the following areas:
> (a) wetlands;
> (b) coastal zones;
> (c) mountain and forest areas;
> (d) nature reserves and parks;
> (e) areas classified or protected under UK legislation, SPAs and SACs;
> (f) areas in which the environmental quality standards laid down in EC legislation have already been exceeded;
> (g) densely populated areas;
> (h) landscapes of historical, cultural or archaeological significance.
>
> **3. Characteristics of the potential impact**
> The potential significant effects of projects must be considered in relation to the criteria set out under 1 and 2 above, and having regard in particular to: the extent of the impact (geographical area and size of the affected population), the transfrontier nature of the impact, the magnitude and complexity of the impact, the probability of the impact, the duration, frequency and reversibility of the impact

Box 13

Government guidance confirms that EIA will be needed if a development is likely to affect SSSIs and internationally important sites (SPA, SAC, Ramsar) [12]. However, as noted above, Government guidance does not give advice as to the weight to be given to BAP targets in the planning process. Although DETR Circular 02/99 does state that *'Where relevant, Local Biodiversity Action Plans will be of assistance in determining the sensitivity of a location'* and thus whether EIA is required (Paragraph 39, DETR, 1999b).

To ensure that road projects having impacts on biodiversity are subjected to EIAs, the assessment of environmental sensitivity in the screening process should include biodiversity criteria to look at what biodiversity elements are likely to be affected (see the Box 14 below). These criteria have been largely derived from the criteria in Annex 3 of the EIA Directive as amended.

---

[12] Circular 02/99 (DETR, 1999b), PPG 9 (DoE, 1994), TAN 5 (Welsh Office, 1996), NPPG 14 and Scottish Circulars 6/1995 and 15/1999 (SOED, 1995 and 1999; Scottish Executive, 1999b), and Development Control Advice Note 10 and PPS2 (DoE – Northern Ireland, 1989, 1997).

Suggested biodiversity screening criteria

(Not mutually exclusive)

Will the road project impact (directly, indirectly or cumulatively) on:

Bioregional/landscape
- The nature conservation characteristics of the bioregional area? Eg Natural Area in England.
- The spatial pattern and connectivity of the habitats in the landscape eg by fragmentation or by leading to extensive edge effects?

Ecosystem/habitat
- An internationally, nationally, regionally or locally designated area? (See Appendix 1)
- An area being officially considered for an international, national, regional or local designation?
- Ancient woodland?
- Non-designated areas of semi natural habitat? Eg see the following box.
- Biological resources (ie genetic resources, organisms or parts thereof, populations, or any other biotic component of ecosystems with actual or potential use or value for humanity) important for the conservation of biodiversity?
- Ecosystems and habitats which are:
   (i) Subject to national, regional and/or local HAPs? (See Appendix 2)
   (ii) Typically associated with species protected under international/national legislation? eg The Birds Directive, the Habitats Directive, the Wildlife and Countryside Act 1981, CITES, etc.
   (iii) Typically associated with species subject to national, regional and/or local SAPs?
   (iv) Required for migratory species?
   (v) Representative of unique biological processes? ie where the processes (eg hydrology, demographic trends) are unique compared to other ecosystems/habitats in the area.
   (vi) Of social, economic, cultural or scientific importance at a national, regional or local level?

Species/communities
- Species and communities which are:
   (i) Protected under international/national legislation? eg The Birds Directive, the Habitats Directive, the Wildlife and Countryside Act 1981, etc.
   (ii) Subject to national, regional, local SAPs or listed in regional or local BAPs? See Appendix 3.
   (iii) Of social, economic, cultural or scientific importance at a national, regional or local level?
   (iv) Migratory
- Attempts to protect ecosystems or promote the recovery of threatened species?

Box 14

Additional guidance is given on the areas of semi-natural habitat to be considered in EIAs (DoT, 1993):

Heathland

> **Areas of semi-natural habitat potentially of biodiversity value**
>
> - Rivers or stream valleys and other wetland areas such as lakes, large ponds, reed beds and gravel pits;
> - Areas of permanent pasture and herb rich meadow;
> - Areas of deciduous and semi-natural coniferous woodland;
> - Farmland with a strong pattern of hedgerows, hill farming and crofting land;
> - Other wildlife corridors such as verges, embankments, old drove roads, disused railways and canals;
> - Lowland heath and scrub;
> - Bogs, mires and fens;
> - Moorland, narrow glens and mountainous areas;
> - Coastal habitats (eg estuaries, dune systems, salt marshes, cliffs and rocky shores);
> - Ecotones ie transition areas where habitat types change from one to another;
> - Derelict areas which have been recolonised by plants and animals.
>
> (DoT, 1993)

Box 15

Where a road project will potentially affect any of the biodiversity elements listed in these screening criteria an EIA should be carried out. The screening criteria explicitly mention wider conservation interests as well as protected areas and species, as the latter should not be the only criteria which trigger and are investigated in road EIAs. As IEA (1995) states:

'Although the identification of designated sites of conservation interests is important for evaluating the baseline environment, care should be taken that an ecological assessment does not place undue emphasis on the presence of these sites at the expense of wider interests. his is because wildlife conservation is reliant upon the protection of the wider countryside in conjunction with a system of individual site designations.'

5. Scoping biodiversity issues

The purpose of scoping is to determine the range of environmental topics (including alternatives) to be addressed, the appropriate level of detail to be applied to each topic area and the methods and approaches to be adopted for their assessment.

The amended EIA Directive includes much stronger scoping provisions. Article 5 and Annex IV of the Directive set out the information that should be contained in an EIS (see Box 16). An EIS *must* contain such of the information referred to in Part I of Box 16 as is *reasonably required* to assess the environmental effects of the development and which the applicant can, having regard to current knowledge and methods of assessment, *reasonably be required* to compile. It *must* include at least the information referred to in Part II. Early consideration of these requirements will be essential during scoping to ensure that the EIA fully complies with the Directive and UK Regulations.

Information for inclusion in EISs

Part I – Include in EIS as far as reasonably required

1. Description of the development, including in particular:
   a) a description of the physical characteristics of the whole development and the land-use requirements during the construction and operational phases;
   b) a description of the main characteristics of the production processes, for instance, nature and quantity of the materials used;
   c) an estimate, by type and quantity, of expected residues and emissions (water, air and soil pollution, noise, vibration, light, heat, radiation, etc.) resulting from the operation of the proposed development.

2. An outline of the main alternatives studied by the applicant or appellant and an indication of the main reasons for his choice, taking into account the environmental effects.

3. A description of the aspects of the environment likely to be significantly affected by the development, including, in particular, population, fauna, flora, soil, water, air, climatic factors, material assets, including the architectural and archaeological heritage, landscape and the inter-relationship between the above factors.

4. A description of the likely significant effects of the development on the environment, which should cover the direct effects and any indirect, secondary, cumulative, short, medium and long-term, permanent and temporary, positive and negative effects of the development arising from:

   (a) the existence of the development;
   (b) the use of natural resources;
   (c) the emission of pollutants, the creation of nuisances and the elimination of waste,

   and the description by the applicant of the forecasting methods used to assess the effects on the environment.

5. A description of the measures envisaged to prevent, reduce and where possible offset any significant adverse effects on the environment.

6. A non-technical summary of the information provided under paragraphs 1 to 5 of this Part.

7. An indication of any difficulties (technical deficiencies or lack of know-how) encountered by the applicant in compiling the required information.

Part II – MUST be included in each EIS

1. A description of the development comprising information on the site, design and size of the development.

2. A description of the measures envisaged in order to avoid, reduce and, if possible, remedy significant adverse effects.

3. The data required to identify and assess the main effects which the development is likely to have on the environment.

4. An outline of the main alternatives studied by the applicant or appellant and an indication of the main reasons for his choice, taking into account the environmental effects.

5. A non-technical summary of the information provided under paragraphs 1 to 4 of this Part.

(CEC, 1985 & 1997; SI 1999 No. 293)

Box 16

The scoping exercise will provide three principle products:

- A list of activities which may cause environmental disturbance, together with initial estimates of their likelihood and of potential magnitude (these should include impacts that may be of concern to the public even if these are for apparently emotional/irrational reasons);
- A list of nature conservation receptors likely to be affected by the project; and
- A plan for conducting the EIA technical studies, including information/data needs, details of methods to be used, the type of magnitude/significance criteria that will be appropriate for assessing impacts, and resources required.

Scoping which receptors should be made the focus of further studies is an essential part of EIA, this is especially crucial in relation to biodiversity where it is simply not possible to measure everything. Hence the need to choose biodiversity receptors rigorously and systematically and to be able to defend the choice of receptors robustly. If receptors have been selected for ad hoc reasons, eg solely because data is available, this could lead to other elements of biodiversity not being chosen as receptors and assessed in the appropriate level of detail which could be very difficult to defend. In turn, this could mean that the project stalls when further survey work needs to be carried out at a later stage.

RPTI (1999) state:

'The effects on biodiversity should be assessed in every statutory environmental statement and considered throughout the environmental assessment process, particularly at the scoping stage. Even where there may be no significant adverse effects on biodiversity the environmental assessment process may highlight opportunities for enhancement. In some cases it may be considered that these help to offset some adverse effects unrelated to wildlife.'

The following sections will guide scoping to ensure that biodiversity issues are fully considered. They provide guidance on the project activities that could lead to impacts, the range of potential biodiversity impacts to consider, the categories of impacts, and the time and spatial parameters for the EIA.

5.1 Project activities

A wide range of activities is potentially associated with road projects and these are summarised in the following box. This checklist can be used to identify which activities will take place for a particular project.

Checklist of activities which may be associated with a road project

Landtake activities	Construction	Operation and maintenance
Pre-site works	Movement of vehicles and plant on and off site	Lighting
Permanent landtake	Presence of people	Maintenance work
Temporary landtake	Earth moving	Vegetation management eg mowing verges
Deep excavations	Storage of oil	
Installation of drainage	Storage of construction materials	Increased use of adjacent areas
Culverting	Accidental spillages	Gaseous emissions
Fencing	Activities causing emissions	Particulate emissions
Erection of structures	Activities generating noise	Noise
Landtake to provide construction materials	Lighting Accommodation and offices	De-icing Lighting
Landtake for induced developments*	Disposal of spoil and waste	Drainage

*Roads may lead to induced effects ie encourage residential, industrial, retail and leisure developments on accessible adjacent land. These effects are potentially very significant and should be considered in the EIA.

Crayfish

Box 17

5.2 Potential impacts

Impacts of roads on biodiversity fall into 4 main types: habitat loss, habitat fragmentation, direct and indirect impacts on habitat quality and species, and cumulative impacts. English Nature guidance (1994a) provides detailed information on impact identification and associated mitigation measures. Spellerberg & Morrison (1998) provide a comprehensive treatment of habitat fragmentation effects. The checklist in Box 18 enables potential impacts for a particular project to be identified.

> **Checklist of impacts to consider**
>
> **Habitat loss effects**
> - Permanent habitat loss on site
> - Temporary habitat loss on site eg land taken up by construction equipment/temporary roads
> - Physical removal of soils and vegetation
>
> **Habitat fragmentation effects**
> - Reduced habitat connectivity in the landscape – can disrupt the established relationships between different habitats or patches of the same habitat eg routes linking sleeping or roosting areas to feeding grounds or migration routes may be physically interrupted.
> - Barrier effects on species – can affect the movement of wildlife: population viability may be affected if populations of a scarce species are separated especially if they have poor dispersal activities
> - Increased mortality due to wildlife casualties
> - Edge effects – if vegetation is removed the new linear gap creates a new microclimate and a change in physical conditions which can extend varying distances from the road edge. This newly created habitat may provide habitat for edge species and facilitate dispersal for some species.
> - Reduced patch size - may reduce populations of key plant species, which in turn may affect the abundance of insects including butterflies they support. These require a minimum area to sustain viable populations and may in turn affect other species eg predatory birds. Also small patch size may not be able to support the range of habitat structure needed to sustain a range of different species
>
> **Changes in habitat quality and other indirect impacts**
>
> *Changes to natural processes*
> - Groundwater regimes - changes in the groundwater regime may adversely affect habitats dependent on the watertable eg marsh, fen and bog. Depending on the geology, lowering the water table can impact habitats a considerable distance from the development.
> - Stream/river flows - Increases or reductions in natural rates of flow eg flash flooding from hard surfaces may affect aquatic ecosystems. Accumulation of construction spoil can alter flow, volume and composition of water. These increased solids increase turbidity which can cause abrasion damage and gill blockage in fish and lead to the disappearance of filter feeding invertebrates
> - Flooding regimes
> - Soil leaching and changes in soil structure
> - Soil erosion patterns
>
> (cont...)

Box 18

Checklist of impacts to consider (continued)

Water pollution
Water pollution from accidental spillages, de-icing chemicals, runoff and road spray can lead to adverse changes in aquatic biodiversity as can changes in sediment and solid loads in watercourses.

Soil pollution
Road spray, vehicle emissions and dust and other particulates (including aggregate and sealant materials used in road construction) can be deposited directly on the land or by polluted precipitation and by polluted groundwater. These can change soil pH and structure. Soil conditions can also greatly alter the effective toxicity of pollutants.

Air pollution
Emissions of lead, zinc, nitrogen, de-icing materials and particulates such as dust can affect biodiversity.

Changes to microclimate
Light and radiation emissions may alter the microclimate. These microclimatic changes may be sufficiently great to alter the performance of some species of plants and animals.

Windfunnelling
Where woodlands are bisected interior trees become exposed and liable to wind-blow effects leading to changes in the new marginal vegetation. Cuttings can have an additional windfunnelling 'jet' effect increasing windblow and evaporation that may result in a water supply shortfall which may lead to changes in species composition.

Disturbance
Fauna can be disturbed by noise, lighting and vibrations from traffic and by road lighting.

Reduced visibility
Road structures eg bridges and viaducts may cause problems for certain birds/mammals by reducing visibility

Introduction of exotics
The edge habitat or ecotone and traffic on the road may facilitate dispersal for some species. This may result in dispersal and establishment of alien and invasive species or pest species that may have secondary effects on biological communities.

Changes to habitat management
eg frequency of verge cutting.

Public pressure
Surrounding habitats may be placed under increasing public pressure, because of access, leading to effects including the disturbance of animals, and physical destruction of ground flora. Also litter may accumulate along roads.

(cont...)

Box 18 cont

> **Checklist of impacts to consider (continued)**
>
> *Off site habitat losses and changes in habitat quality*
> In relation to the obtaining and disposal of materials eg mining for aggregates for road building.
>
> **Cumulative effects**
> Even relatively minor habitat loss, fragmentation and indirect impacts of an individual road project can, when added to other past, present and reasonably foreseeable future impacts of other projects and activities, contribute to significant impacts in an area. All relevant types of future projects and activities should be considered (ie not just other road projects) including induced development.
>
> **Positive effects**
> - Habitat enhancement
> - Improved habitat management
> - New structures eg bridges and tunnels may provide habitats for some species eg bats
> - Habitat creation
>
> (I. Spellerberg, personal communication)

Box 18 cont

As noted above, one of the key outputs of scoping is to identify the list of biodiversity receptors that potentially may be affected. The checklist in Box 19 can help the systematic identification of receptors in a particular case. It is recognised that detailed studies at a genetic level are unlikely to be appropriate for the majority of EIAs. Collection of detailed genetic information is unlikely to be considered reasonable (having regard to current knowledge and methods of assessment, etc. (see Box 16)) for specific projects at the present time. However, genetic level receptors should be identified, as even in the absence of detailed information, the precautionary principle should be applied to ensure the protection of valued elements. The issue of genetic studies is discussed further in section 6.

Biodiversity Impact

> ## Checklist of biodiversity elements to consider
>
> *Bioregional level*
> - The nature conservation characteristics of the bioregional area (eg English Nature natural area in England) and designated sites
>
> *Landscape*
> - The spatial pattern of all the habitats in the landscape
> - Connectivity of habitats including potential wildlife corridors
> - Opportunities for habitat creation/enhancement
>
> *Ecosystem/habitat/community levels*
> - All habitats and communities in the area including priority and BAP/LBAP habitats and species
>
> *Species level*
> - Endangered/threatened species
> - Endemic species
> - Protected species
> - SAP species
> - Characteristic species of each habitat
> - Species with low reproductive capacity, eg most large mammals
> - Species highly sensitive to disturbance eg most birds of prey
> - Species subject to recovery programmes
>
> *Population level*
> - Populations at low levels in cycle, eg salmon stocks in some rivers;
> - Populations at outer limits of their range
> - Declining populations
> - Metapopulations
>
> *Genetic level*
> - Genomes and genes of social, scientific or economic importance eg agricultural crops, domesticated species
> - Isolated populations
> - Genetic diversity/phenotype

Box 19

5.3 Categories of impacts

Types of effects to be assessed should include direct, indirect, secondary, cumulative, short, medium and long-term, permanent and temporary, positive and negative effects of the project. In relation to biodiversity it is particularly important to consider indirect and cumulative effects as well as direct effects.

Kestrel

> **Cumulative environmental effects**
>
> Cumulative effects are effects on the environment that are caused by a project in combination with those of other past, present and future projects and activities.
>
> In practice assessing cumulative effects requires an EIA to:
>
> - Assess effects over a larger (ie regional) area
> - Assess effects during a longer period of time into the past and future
> - Consider effects on receptors due to interactions with other projects and activities, not just the effects of the project under review
> - Include other past, existing and future (eg reasonably foreseeable) projects and activities
> - Evaluate significance in terms of different spatial and temporal scales ie consideration of other than just local, direct effects.
>
> Aspects of the assessment of cumulative effects on biodiversity are discussed in later sections of this guide. A list of useful general references on cumulative effects assessment is included in Appendix 4.

Box 20

5.4 Spatial/time issues

Time and spatial parameters of the study are defined in the scoping stage and it is vitally important to the long-term viability of biodiversity that these definitions consider ecological processes and components such as migratory or nesting patterns for birds.

Appropriate boundaries are crucial for considering biodiversity in EIAs and broader spatial scales and longer time scales are needed than those traditionally used in ecological impact assessments (IAIA 1999 Biodiversity Working Group, unpublished). Therefore, it is important to examine the proposal not only for effects at the local level but also for effects at the larger, bioregional ecosystem level. Evaluating the proposal within a larger bio-regional/landscape-level context will ensure that a variety of local and regional biodiversity concerns, including cumulative effects, are addressed. The analysis of effects should cover the largest relevant scale (based on the affected resources and anticipated effects) as well as local scales.

> **Guiding Principle**
>
> The study area should reflect the impact type (eg indirect effects will often extend throughout a watershed) rather than taking a fixed width corridor approach

Box 21

Ideally the study area for each impact should reflect the area likely to be affected by that particular impact type. For example, a relatively large area will need to be studied for potential hydrological impacts whereas the area studied for potential effects of road lighting is likely to be more restricted (Forman & Deblinger, 1998).

Another approach is to set appropriate temporal and spatial boundaries for each biodiversity receptor and for the spatial boundaries to reflect the distribution and patterns of movement of a particular receptor. For example, the boundaries for migratory bird populations may extend beyond the traditional project study area because deterioration or loss of breeding habitat could influence population levels and resource use over extensive areas (eg regional, national, international areas). The following approach illustrates some of these considerations.

Table 3 - Setting boundaries

Receptor	Temporal boundaries	Spatial boundaries
A specific plant species	Year round	A specific designated area
Aquatic birds	Year round April - July	Wider boundary: Europe Immediate boundary: specific habitat areas near the proposed project location
Bird movement and migration	Year round Spring and Autumn	Wider boundary: Europe Immediate boundary: 2 km corridor around the proposed project location
Terrestrial birds	Spring and Summer Spring and Summer	Wider boundary: Europe Immediate boundary: 500m corridor around the proposed project location
Freshwater resources	Year round	Close proximity to the project corridor
Groundwater resources	Year round	Close proximity to the project corridor

(Based on the boundaries used in a Canadian road bridge EIS (Jacques Whitford Environment Limited, 1993)).

World Bank Guidance (1997) advocates production of a scoping map which gives a total picture of the project site and the areas likely to be affected by the different types of impacts during the different stages of the project.

DoE Guidance (1995) stresses the need for a view to be taken at the outset of the timetable for completing the EIA *'For a major EA it may run for 12-18 months from the decision to initiate the EA to production of the final ES. This reflects a situation where background information on for example, flora and fauna, climatic conditions and noise and dust, may need to be collected over a full year in order to identify seasonal variations.'*

In addition, time will need to be allocated to investigating existing data sources to provide historical trends and background information.

5.5 Consultees

Developers proposing road projects should have full and early consultation with the relevant bodies. These will include the planning authority, statutory nature conservation consultees (English Nature, Countryside Council for Wales, Scottish Natural Heritage, Northern Ireland Environmental Service: Countryside and Wildlife), all the usual nature conservation consultees (eg Wildlife Trusts, RSPB, specialist local groups), the relevant regional and/or local biodiversity partnership/initiative groups[13], all other bodies which have an interest in the likely biodiversity impacts of the project (eg the Environment Agency, the Scottish Environmental Protection Agency) and local communities. These parties should be invited to participate in the scoping process. Consultation with these groups will ensure that potential issues are not missed leading to costly work at a later stage, they can also identify additional sources of data or information, and eliminate consideration of unnecessary impacts.

---

[13] See the UK Biodiversity Secretariat's list of LBAPs (DETR, 1999a, also available on the Secretariat's webpage at http://www.jncc.gov.uk/ukbg) for the relevant contact(s).

5.6 Scoping outputs

The findings of the scoping process may be formally presented in the form of a Scoping Report, although the production of such a report is not a requirement of UK EIA Regulations. A scoping report can provide a developer with a valuable check on the progress and competence of the EIA team and provide an opportunity for interested parties/experts to comment on the proposed coverage and methodology of the EIA (English Nature, 1994b).

Circulation of a short scoping report summarising the proposed biodiversity (and other) receptors and the scope of the proposed study to all the statutory consultees and other key organisations for agreement can be extremely useful. Achieving consensus at this stage can avoid delays due to objections as to the adequacy of the EIS at a later stage (IEA, 1995).

6. Baseline conditions

The baseline biodiversity conditions in the area of the proposed road project need to be described. This description will be based on the information provided by consultees, background sources of information and the results of new surveys carried out for the EIA. A summary of the baseline information which is needed is given at 6.4. The description of baseline biodiversity conditions is vitally important for subsequent stages of the EIA:

'A prediction of change is only as effective as the baseline information. It is not possible to attempt to assess the predicted effects of a proposed development unless the existing conditions are clearly and accurately recorded, presented and understood' (RSPB, 1995).

Good sources of background biodiversity information and some key scientific references are available and should be used to supplement usual sources of ecological/nature conservation data (see Reference Boxes 3, 4 & 5).

6.1 Consultees

The consultees who were involved in the scoping process (see 5.5 above) (and any others subsequently identified) should be asked for any relevant information.

6.2 Relevant information on other projects/activities

The consideration of cumulative impacts will require collation and analysis of information on past, existing and future projects and activities. Possible sources of existing information may include: DETR, Highways Agency, Scottish Executive, Welsh Assembly, consultees, local planning authorities, project developers, promoters and operators, local academic and research institutions, local residents and community and environment groups. For past and existing projects an EIS may be available which will provide a good source of information. It will only be feasible to consider future projects and activities that are reasonably foreseeable – the bodies involved should be contacted for information about these proposals.

6.3 New surveys

Nearly all road EIAs will include some new biodiversity survey work. The new survey work should generate data on the status of biodiversity at each of the appropriate levels, sufficient to make defensible and robust impact predictions (World Bank, 1997; Bagri, 1998).

As biodiversity encompasses variability at various different levels there are many different measures of biodiversity. Options for measuring biodiversity include: measuring species richness, family richness, species abundance, phylogenetic measures (which measure how closely related in evolutionary terms species are and tend to capture not only the degree of relationship, but also the degree of difference in many other characteristics (Gaston & Spicer, 1998)), taxonomic measures, molecular measures, the presence of certain species, diversity indices, and biodiversity indicators. Some of the key measurement references are detailed in Reference Box 6.

For EIAs, typically with time and resource constraints, the key issue is to ensure that the data collected are relevant ie that appropriate data are collected to answer clearly defined questions. The traditional UK approach to ecology surveys for EIA – where often very little data is collected, in many cases only a phase 1 habitat survey (JNCC, 1993) with species lists and presence/absence records for protected species - *will not* provide appropriate data for a rigorous biodiversity assessment. Very little abundance data is generally collected for EIAs, but without this it is extremely difficult to assess the significance of likely impacts on populations.

'Biodiversity specialists working on [EIAs] have a responsibility to ensure that they exercise best professional judgement as to the minimum data needed to characterise the environment and to make defensible impact predictions. The key challenge is to produce a sufficiently detailed impact analysis in the face of: insufficient data; inadequate knowledge of the affected ecosystem(s), habitat(s), or species; and uncertainties over cumulative impacts' (World Bank, 1997).

Another option is to use biodiversity indicators if available[14]. These would facilitate less cost intensive assessments of biodiversity for use in EIAs. However, detailed analysis of the effectiveness of proposed indicators is at an early stage. Until this work has been completed the use of biodiversity indicators should be treated with caution (eg Prendergast & Eversham, 1997).

In the meantime, EIA consultants are left with the issue of how to focus new survey work in order to collect the most meaningful biodiversity information. To help this process detailed interviews were carried out with a range of interviewees to discuss which biodiversity measurements were felt to be the most relevant for EIAs. The measurements considered to be most useful by interviewees were incorporated into the Biodiversity Information Framework set out in section 6.4.

The minimum new survey requirements needed are summarised in the following Box. The focus of these is to summarise the biodiversity information set out in the suggested Biodiversity Information Framework in section 6.4.

Minimum new survey requirements

- A survey of all the habitats in the area likely to be affected. This should include an assessment of the quality of each habitat (broadly equivalent to a Phases 1 and 2 Habitat Survey).

- More detailed survey work (determining species abundance and distribution) on selected key species. The key issue here is the correct focusing of the species for detailed study – see sections 6.3.2 - 6.3.4 below

---

Box 22

Several of the key issues arising from these measurements/indicators are discussed in more detail below.

6.3.1 Ecosystem/habitat quality/'biodiversity potential'

Biodiversity status is not simply about what ecosystems/habitats are present, but also their quality. As noted above, the habitat survey must assess the quality of each habitat type. The JNCC are producing a detailed interpretation manual on HSs and when this is available it will probably be useful for the habitat surveys. To a large extent the EIA needs to focus on habitats outside designated areas (especially SSSIs). Habitats in designated areas have already been assessed as being of relatively high quality and current government policy is that wherever possible such sites should not be damaged by roadbuilding (DETR, 1998b), though in practice this is not always the case. Additionally, it is vital to consider the biodiversity value of the wider

---

[14] Reid *et al*, 1993; Noss, 1990; UNEP, 1997a,b and c are just a selection and English Nature are in the process of developing an indicator approach for their common standards monitoring. The expert advisory body of the CBD (the Subsidiary Body for Scientific, Technical, and Technological Advice (SBSTTA)) is working towards a core set of biodiversity indicators with the aim that a 'first track' set of indicators are available by 2000 (UNEP, 1997a and b).

countryside as the designated site network will not in itself protect UK biodiversity. The following table sets out some useful criteria for assessing habitat quality.

Table 4 - Criteria for assessing habitat quality

Criteria	Key references
Ratcliffe criteria Used as the basis of site designation Volume 11 (DoT, 1993) suggests that these are used as the basis for assessments of general wildlife value	Ratcliffe, 1977
Favourable Conservation Status (FCS) and Integrity These concepts were introduced by the Habitats Directive for Natura 2000 sites, but could equally be applied to other designated sites and the wider countryside	Habitats Directive – see Annex 1 ALGE *et al*, 2000 PPG 9 (DoE, 1994) (see Appendix 1) defines 'site integrity' Welsh Office, 1996 SOED, 1995 DoE – Northern Ireland, 1997
Common Standards Monitoring The Countryside Council for Wales, English Nature and Scottish Natural Heritage are jointly establishing common standards throughout Great Britain for the monitoring of nature conservation. Although the work so far focuses on assessment of the condition of designated sites the suggested framework could be applied more widely	JNCC, 1998
EIA Habitat Based Methods In the US several habitat-based systems have been developed. These systems use habitat characteristics to infer whether or not a habitat is of good quality in terms of its ability to support appropriate animal and plant species. The quantitative use of such methods requires considerable information and the development of numerical indices of habitat quality and is therefore not likely to be practicable, but the methods could be used on a qualitative basis.	Canter, 1996 chapter 11 provides a good overview US Army Corp. of Engineers (1990) A Habitat Evaluation System for Water Resources Planning, Lower Mississippi Valley Division, Vicksburg, Mississippi US Fish and Wildlife Service (1980) Habitat Evaluation Procedures (HEP), ESM 102, US Fish and Wildlife Service, Washington, DC, Mar.

The habitat quality assessment should also identify areas that have potential for biodiversity enhancement measures. If the key objective of biodiversity assessment as set out in this guidance is to be achieved, schemes will routinely need to incorporate biodiversity enhancement measures. Therefore, areas with potential for biodiversity enhancement need to be identified as early as possible in the EIA process so that there is scope for thorough planning of enhancement measures.

6.3.2 Key species groups

'The inability to cover all groups of organisms constitutes another area in which all ESs are potentially open to criticism and demands for extra work' (English Nature 1994b). Because of this, determination of the key species on which detailed survey work will be carried out is fundamentally important. The key species will generally cover a range of species groups eg plants, birds, mammals, amphibians, reptiles, invertebrates. The reasons for selecting the chosen species must be defensible – not just because the species are the easiest to survey!

Traditionally UK EIAs have concentrated (understandably) on protected species, but other species eg locally important species should also be included. Key species will generally be species in at least one of the following categories (which are often identified in LBAPs):

- Threatened species
- Endemic species
- Protected species
- SAP species (national, regional, or local)
- Characteristic species for each habitat
- Species susceptible to habitat fragmentation or disturbance

Typically, a relatively small number of key species will be chosen – the selection process should involve participation of consultees, not be made by the EIA consultant in isolation.

Pied flycatcher

6.3.3 Characteristic species

Characteristic species are species usually associated with a particular habitat eg dragonflies in lowland rivers, craneflies in alder woods, mosses and lichens in lowland bogs, redstarts, wood warblers and pied flycatcher in upland oak woodland, and river water crowfoot, starworts and water cress in chalk rivers. They are not necessarily rare and assessing their status (ie population levels and distribution) can be useful in establishing the status of the associated habitat ie the 'quality' of the habitat can be inferred from the status of the populations of characteristic species.

6.3.4 Species susceptible to habitat fragmentation

All species are potentially susceptible to habitat fragmentation (decreasing size of habitat patches or changes in size distributions) to some extent. However, some are more susceptible eg top predators (raptors, carnivores), species with limited dispersal, many small mammals (see English Nature, 1994a for a more detailed discussion), species with cyclic populations and short or non-overlapping generations (eg annual plants), species with complex life-histories (eg amphibians, many insects) and specialist mutualists (eg pollinators, symbionts). The species in the study area that are likely to be most susceptible should be identified and it may be appropriate to include these in the list of key species. Appendix 3 notes the national SAP species that are threatened as a result of habitat fragmentation species eg red squirrel and sand lizard.

6.3.5 Diversity indices

Discussions on biodiversity often mention species diversity indices of which there are many. Some are based only on the number of species present and the species composition (species richness indices); others take species abundance into account (diversity indices). Sometimes indices are useful for EIAs: BMWP (Biological Monitoring Working Party) and ASPT (Average Score Per Taxon) indices are commonly used in EISs to present information about freshwater invertebrates. Indices are of most use for comparing different areas of the same habitat type. Where used in the main volume of the EIS it is essential that diversity indices are explained so that they are understandable by the non-ecologist.

6.4 Biodiversity Information Framework

A framework for the information/measurements needed for each level of organisation is set out in Table 5. While it is recognised that any particular EIS is unlikely to present all of this information, use of this framework to structure the collection of biodiversity information should help ensure that biodiversity issues are given proper consideration based on appropriate and adequate information.

Table 5 - Biodiversity Information Framework

Biodiversity level	Information which EIS should include	Information/measurements	Why this information/ measurement should be included
Biogeographic Area	Characteristics of and objectives for Biogeographic Area	*Composition* In England information from Natural Areas CD-ROM (English Nature 1998d) and any updates especially: • Characteristics of Natural Area (NA) • Objectives for NA • Regional biodiversity indicators. • Broad scale biodiversity information from any relevant multi-modal transport/corridor studies *Structural* • Connectivity • Spatial linkage • Patchiness • Fragmentation • Configuration • Juxtaposition *Functional* • Disturbance processes • Nutrient cycling rates • Energy flow rates • Hydrologic processes • Landuse trends	*Overall rationale: to provide the regional context for the EIA* *Composition* • To give an indication of the relative abundance/rarity of the different habitats/species in the bioregional area • The NA objectives/regional biodiversity indicators will help set 'assessment criteria' for determining impact magnitude/ significance at a regional scale for a particular project *Structural* To give an overview of: • The spatial pattern of elements in the landscape • The quality of the different habitats within the natural area • The types of species that are likely to be present *Functional* To give an overview of: • Past and present management regimes in the area • The species that are likely to be present
Landscape	Pattern of elements in the landscape	*Composition* • Landuses (current and historical) • Number, identity (diversity) and distribution of habitats in landscape • Presence of biodiversity 'hotspots' • Ecosystem boundaries (eg watersheds) • Potential wildlife corridors • Area to be lost in landtake to scheme • Total area likely to be affected by scheme • Total area / % of natural/ semi-natural habitat to be lost in landtake to scheme • Total area / % of natural/ semi-natural habitat likely to be affected by scheme *Structural* • Connectivity • Spatial linkage	*Overall rationale: to provide the detailed context for the EIA* *Composition* • To provide details of the areas of different habitats in the landscape, their relative abundance/rarity and pattern which will help to set meaningful criteria for determining impact magnitude/ significance at a regional level for a particular project • Quantification of areas of different habitat lost is necessary to allow impact magnitude to be determined • Consideration of past losses (cumulative effects) and their implications for current and future availability of habitat at the landscape level

		• Patchiness • Fragmentation • Configuration • Juxtaposition *Functional* • Disturbance processes • Nutrient cycling rates • Energy flow rates • Hydrologic processes • Landuse trends	*Structural* To provide more details about: • The spatial pattern of elements in the landscape and the likely ability of species to move between them in a reconfigured landscape • The quality of the different habitats within the natural area • The likelihood of edge effects • The types of species that are likely to be present – ie what species are characteristic of each habitat and the implications of the particular pattern of habitats in the landscape for these *Functional* To provide more details on: • Past and present management regimes in the area • The likely effects of landscape reconfiguration on the species that are likely to be present • Changed availability of habitat in relation to species mobility • Impacts of landscape reconfiguration on energy flow and carrying capacity • Possible longer term impacts of land use change
Ecosystem/ Habitat/ Community	Quantity and quality of each ecosystem/ habitat/ community	*Relevant HAPs* • National, regional local HAPs applying to habitats in the area *NVC classification* • Identifies and describes vegetation communities in areas to be lost/affected *Quantity of habitat lost/affected* • Area /% of each habitat to be lost to the scheme in land take • Area /% of each habitat likely to be affected by specific impacts of the scheme eg large area of habitat may be affected by indirect impacts from hydrological changes, a smaller area by traffic noise • Fragmentation of ecosystems/habitats caused by the project • Links between habitats that would be altered by the project	*Relevant HAPs* • Targets in HAPs will help set 'assessment criteria' for determining impact magnitude/significance at a local level for a particular project • Knowing where habitats with HAPs are will help the evaluation of biodiversity 'value' of the areas to be affected by the project. These habitats should be avoided *NVC classification* • Enables local losses to be evaluated in a national context • Identifies key species (eg protected, SAP, locally important) in the area which may require further study *Quantity of habitat lost/affected* • Allows impact magnitude to be quantified • Knowing the areas likely to be affected by specific

		Quality of each habitat lost/affected Assessment of quality of each habitat: • Designated areas • Areas with HAPs • Areas with potential for biodiversity enhancement • For habitats which will be lost detail areas/% that are: in designated sites, subject to HAPs, of 'good quality', have potential for biodiversity enhancement • For habitats likely to be affected detail areas/% that are: in designated sites, subject to HAPs, of 'good quality', have potential for biodiversity enhancement • Susceptibility of each habitat to edge effects especially the degree of transformation and the impacts of pollution from the road *Structural* • Substrate and soil variables • Slope and aspect • Foliage density and layering • Abundance, density and distribution of key physical features eg out crops • Water and resource availability *Functional* • Resource productivity • Population dynamics (including metapopulations) are relevant at the ecosystem level – these are dealt with separately - see the population level information/measurements below • Predation rates • Patch dynamics • Nutrient cycling rates • Human intrusion rates and intensities	types of impacts ensures that detailed studies are carried out in appropriate locations and enables appropriate avoidance/mitigation/ compensation measures to be planned *Quality of each habitat lost/affected* • This enables prioritisation of all the habitats in the area to be affected ie which must be avoided, where mitigation/compensation /enhancement might be appropriate and feasible • Looking at the range of habitats (ie designated/those with HAPs, etc.) avoids the common overemphasis on nationally protected areas • Details of the actual areas lost enables more precise expressions of impact magnitude/significance *Structural* • Provide information for the assessment of habitat quality/suitability for key species *Functional* • Provide information for the assessment of habitat quality/suitability for key species • Consideration of human intrusion will be useful for assessing how resilient habitats are to current levels of impact and will help plan feasible mitigation measures/future management strategies
Species	Key species	*Composition* Set out the key species and explain why each of these has been selected. Typically these species will include: • Threatened species • Endemic species • Protected species • SAP species (national, regional, local) • Characteristic species for each habitat • Species particularly sensitive to loss, habitat fragmentation and degradation or to other impacts identified *Structural* • Dispersion • Range • Habitat availability	*Composition* • Explicit statements of why each key species has been chosen for study will demonstrate the systematic process used to choose these and make the exclusion of other species more easily defensible • Looking at the range of suggested key species (eg threatened, protected, SAP, characteristic, etc.) should enable declining trends in species which are not currently threatened to be picked up • SAP/HAP/LBAP targets will help set 'assessment criteria' for determining impact magnitude/ significance at a local level

		• Population structure eg sex and age ratios *Functional* • Demographic processes eg fertility, survivorship, mortality • Metapopulation dynamics • Population fluctuations	for a particular project *Structural and functional* This information enables the species distribution and abundance trends collected (see below) to be put in context, particularly with respect to habitat availability
Populations	Population status of key species	*Composition* • Species distribution trends for key species • Species abundance trends for key species *Structural* • Dispersion • Range • Population structure eg sex and age ratios *Functional* • Demographic processes eg fertility, survivorship, mortality • Metapopulation dynamics • Population fluctuations	*Composition* This information enables: • The viability of the populations of key species in the area to be assessed • The natural variation in the absence of the project to be assessed *Structural and functional* This information enables: • Assessment of the viability of populations in the area • The natural variation in the absence of the project to be assessed • Appropriate mitigation measures to be planned eg to ensure that mitigation measures such as toad tunnels are placed in the optimum locations
Individuals and genes	Generally it is not feasible to try to take direct measurements of genetic resources for EIAs – *'...genetic diversity requires considerable information and may be impracticable to apply'* (CEAA, 1996a). Where there are isolated populations detailed work may be appropriate but generally indirect methods will be sufficient	*Composition* • Consider gene flow – eg in relation to connectivity of populations in the landscape • Identify isolated populations that may potentially subject to inbreeding depression *Structural* • Census and effective population size • Genetic diversity *Functional* • Inbreeding depression • Outbreeding rate • Gene flow	• Consideration of connectivity/ identification of isolated populations enables appropriate avoidance/ mitigation measures to be planned/ based on the precautionary principle • All of the other suggested measurements/ information help population viability of particular key species to be assessed in more detail Note: it is acknowledged that some species tend to have low genetic variability naturally, but where there is any uncertainty as to whether a species is at risk from population isolation/ inbreeding depression, the precautionary principle should be applied.

Notes:

1. *A key issue is how the information/measurements collected should be used for evaluation. At present we do not know enough about many of the parameters for particular habitats and species to formulate appropriate evaluation levels. For example, there is a serious lack of safety margins (eg the level of connectivity needed to maintain viable populations of particular species) to enable meaningful evaluation of potential impacts in a particular case (eg whether the fragmentation of a habitat resulting from the road proposal will affect the population viability of a specific population). In these circumstances a precautionary approach must be adopted.*
2. *It will not be possible to measure some of the parameters listed in the structural and functional relationships column in the context of an EIA eg nutrient cycling rates, energy flow rates, but these factors should be considered as far as possible in the survey work and at least addressed qualitatively when quantitative measures are not possible.*

The biodiversity information framework facilitates efficient planning of survey work and data collation. This systematic approach ensures that adequate and appropriate information is collected to enable full consideration of the potential biodiversity impacts.

A summary of how this proposed biodiversity survey approach is different to the surveys traditionally carried out for EISs is set out in Box 23 below.

Differences from 'traditional' EIA surveys

The development and its impacts are viewed from the perspectives of the different biological units and taking an integrated view of the impacts on the affected units (eg not treating birds and terrestrial communities as unrelated entities) which attempts to assemble the whole picture (across scales) with the network of interactions and interdependencies. This involves:

- Explicit treatment of various levels of biodiversity.
- Setting the survey in the wider context of the relevant biogeographical area(s).
- Study areas reflecting areas likely to be affected by different impact types.
- Specific consideration of structural and functional relationships.
- Explicit assessment of habitat quality.
- Key species cover a range of species not just the rare species.
- More detailed abundance and distribution data collected on key species.
- Interface with the UK biodiversity processes via HAPs and SAPs.

Box 23

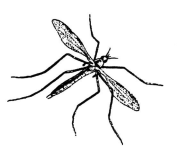

Cranefly

The new surveys must be carried out using good survey practice ie as to timing, study area, methodologies, repeated sampling, etc. and these details must be recorded in the EIS. Although there is much good guidance on this [15] this aspect of EIAs is often weak[16]. A checklist of good survey practice is set out in the following box. The IEA (1995) reference is particularly comprehensive. Information collected at this stage should be made widely available eg to local communities and biodiversity information networks.

---

[15] For example, IEA, 1995; Morris & Therivel, 1995; English Nature, 1994a - see Reference Box 2.
[16] Treweek *et al*, 1993; Thompson *et al*, 1997; Byron *et al*, 2000 - Reference Box 5.

Good survey practice checklist

- Assess the scale of the whole area surrounding the potential development site for which the impact is to be assessed ie the study area for each impact type should be an ecologically meaningful unit. For example, if a road is to pass through a heathland, the ecological survey should consider the entire heathland rather than just the road route, and impacts on hydrology should be considered in the context of the appropriate watershed rather than a fixed corridor.
- Decide on the main season or whether through the year surveys are required, considering the importance of the site in migration, breeding and winter periods.
- Assess the level of normal variation to be expected. Does the population of some species fluctuate wildly? Are key species elusive? Decide the number and spread of visits eg generally 5-10 visits are required to adequately detect birds from a range of habitats in temperate climates, with five being the minimum.
- Decide whether to conduct a full survey or a sample survey. If the area is especially large a sample survey may be better unless the area is composed of a mosaic of different habitats, in which case, a stratified design would be more appropriate.
- Decide on the specific method – whole wildlife community or is the location of key species more important. If a compromise is promoted usually neither the common species nor the rarer ones will be adequately covered.
- Assess the skills of the observer/recorder and use trained staff. It is important that professionals who understand the process of interpretation are employed. Specialist amateurs may need to be approached to cover some difficult groups (eg bryophytes).
- Decide early on the methods to be used to capture data (maps, recording forms, etc) and how it will be computerised.
- Determine the best methods of outputting the data eg mean monthly maxima, graphically presented, in map form, as territory clusters, as mean and maximum counts per season, as bird-days, and/or as an index. This will often be species specific.
- Decide whether and how to incorporate existing data from a diversity of sources. Provide an assessment of its strengths/limitations for EIA purposes, recognising that much of the data will have been collected for different objectives and so is unlikely to be available in a standard format.
- Decide on the method of conservation evaluation (see 6.5 below).

(D.Hill, Personal communication)

Box 24

6.5 Evaluation criteria: Assessing the importance of biodiversity elements

The Government guidance for road EIAs - Volume 11 (DoT, 1993) requires EISs to state the criteria used to evaluate the importance of nature conservation elements, but in practice this is often omitted (Byron *et al*, 2000). Established evaluation criteria are discussed in English Nature (1994b) and DoE (1995) and are summarised below.

Table 6 - Evaluation criteria

Element	Criteria	References
Sites and habitats	Scientific interest	• Ratcliffe (1977) • Usher (1986) • Nature Conservancy Council (1989)
	Ancient countryside importance	Rackham (1986)
	Urban and amenity criteria	English Nature (1994b)
	Geographical level of importance eg International, national, regional, district or local	DoE (1995)
Species	Rarity and vulnerability	English Nature (1994b)
	Ecosystem importance	English Nature (1994b)
	Heritage value	English Nature (1994b)

In practice, EISs use a range of evaluation criteria. Some good practice examples of criteria used in recent road EISs are shown in Table 7.

Table 7 - Examples of evaluation criteria used in recent road EISs

EIS and type of scheme	Criteria/Categories
A249 Iwade Bypass to Queenborough Improvement (Highways Agency, 1997) (A 5.3 km off-line improvement to 2 lane dual carriageway with a bridge)	**Habitats** • First considers the legal or quasi-legal status of sites then suggests that resources can be further defined by use of non-statutory criteria (Ratcliffe, 1977; NCC, 1989); also whether habitat is irreplaceable or could be created elsewhere. • Categories used: *International*: SPA, Ramsar, SACs, pSPAs, and cSACs *National*: NNRs, SSSIs, Environmentally Sensitive Areas (ESAs) *County*: Sites of Nature Conservation Interest (SNCI) **Species** *International importance*: as defined by the Birds or Habitats Directives or listed in the Red Data Book or RDB List. Also species recognised as internationally important under established criteria eg that adopted by the Ramsar Convention *National importance*: Nationally scarce or rare as specified in the RDB *Local importance*: County scarce or rare

A452 Leamington-Kenilworth Road Improvement Scheme (Warwickshire County Council, 1994) and A429 Barford Bypass (Warwickshire County Council, 1996) (An improvement of a 1.6 mile length of single carriageway with poor alignment and substandard width to dual carriageway standard and a two-way single carriageway bypass of approx. 1.2 miles, respectively)	Uses the Ratcliffe criteria and the Warwickshire Biological Records Centre 'Eco-grade' system. Grades given for importance of receptor: *National and Regional*: NNRs, LNRs, SSSIs and sites of regional importance *County*: Prime sites for wildlife, nature reserve potential *District & Parish*: varies from site of ecological importance to good but damaged ecology *Residual*: damaged or impoverished wildlife *None*: little wildlife survives *Uncertain*:???

M25 Motorway Link Roads between Junctions 12 and 15 (DoT, 1994) (A proposal to increase the capacity of the M25 between junction 12 (M3) to junction 15 (M4) by the addition of 2 or 3 lane link roads parallel and on either side of the existing M25 over a distance of approx. 7 miles. Also to widen the existing M25 through junctions 13, 14 and 15 and re-routing of part of the A30)	**Habitats and/or sites** - Used 'the standard criteria' (Ratcliffe, 1977; NCC, 1989) - Categories: *International importance*: Internationally designated sites such as Ramsar sites, SPAs, Important Bird Areas (IBAs) *National Importance*: SSSIs, NNRs and other nationally outstanding sites *County importance*: LNRs, SNCIs or their equivalent, and other sites which are of particular interest or have few or no other examples in the county *Local importance*: semi-natural sites which occur elsewhere in the county supporting a good range of wildlife species or local specialities; due to the variable nature of such sites, they are sometimes qualified as of high local value or low local value *Low interest and very low interest*: these tend to be highly modified, species poor sites such as arable fields **Criteria for birds** - For evaluation of international importance follow the Ramsar Convention. - Also used annually published qualifying levels for national and international importance - Assemblages of breeding birds were assessed using criteria established for evaluation of sites of national importance (NCC, 1989) - Presence of species listed in Annex 1 Birds Directive - Presence of species listed in Schedule 1 WCA 1981 - '1% of population' threshold (this defines a significant proportion of the total population) - Presence of RDB species and candidate RDB species - For rare species, numbers of 100 at a site are considered to be of importance on the International scale, and 50 on the National scale **Criteria for other species** - 'Relevant sources such as the British Red Data Books and other established texts have been used as appropriate when assessing the rarity of species.'

Each of these examples bases nature conservation evaluation on the level of importance of each site/habitat, but there is a lack of consistency in the emphasis given to sites/habitats of local importance. Also the criteria do not link with relevant HAPs/SAPs/LBAPs and consequently risk missing priority species or habitats which are not covered by designations. It can be seen, therefore, that current good practice still falls short of the assessment needed to fulfil the requirements of this guidance.

DETR have recently formulated a New Approach to Appraisal (NATA), which includes evaluation categories (Tables 8 and 10).[17] These criteria could also be applied during EIA when a more detailed level of information is available, as opposed to the (usually) limited information available for an investment appraisal.

Ancient woodland

The DETR publication *Guidance on the New Approach To Appraisal* (DETR, 1998c) discusses NATA. It suggests that sites/features are classified into one of five categories (A to E, with A being of highest importance) and then sets out a basic evaluation methodology (summarised in Table 13 below). The factors on which the evaluation is made are:

- Scale at which the feature matters (eg international, national, regional or local)
- Importance of a feature (eg the reasons why a site was designated)
- Rarity (trend in relation to target): The abundance of the habitat/feature relative to its target level (where appropriate) and its trend, where known, (eg in relation to BAP targets)
- Substitution possibilities: A judgement on whether the habitat/species are substitutable.

However, it is noted that *'some flexibility may be needed in judging the nature conservation evaluation. For example, it may be considered that a site is not designated but could still be important, as the SSSI series is representative, but not all-inclusive. Conversely, a site hosting a single individual of a widespread Berne Convention species may not warrant the highest classification...'* (DETR, 1998c).

---

[17] NATA was initially developed to provide 'a clear and open framework to appraise and inform the prioritisation of trunk road investment proposals' (DETR, 1998c and d). Further work has subsequently been carried out to develop NATA into a multi-modal environmental appraisal methodology for transport planning in general. This multi-modal methodology has just been published as *Guidance on the Methodology for Multi-Modal Studies (GOMMMS)* (DETR, 2000). The intention is that GOMMMS will ultimately form the basis for all intermodal and single mode transport appraisals. However, pending publication of supplementary GOMMMS guidance (expected later this year), the approach in NATA remains the relevant method for appraising road investment (both for the Highways Agency and local authorities).

Table 8 - NATA Evaluation categories

Category	Relevant sites
A	• Ramsar Sites • World Heritage Sites (Convention for the Protection of World Cultural & Natural Heritage, 1972) • Biosphere Reserves (UNESCO Man & Biosphere Programme) • European Sites: Special Areas of Conservation (SACs), Special Protection Areas (SPAs), Sites of Community Importance (SCI), candidate SACs (cSACs) and potential SPAs (pSPAs) • Sites hosting habitats/species of EC importance (Annexes 1 and 2 of the Habitats Directive • Sites hosting species listed under the Bonn Convention (Convention on the Conservation of Migratory Species of Wild Animals) • Sites hosting species listed under the Berne Convention (Annex 1 and 2 of the Convention on the Conservation of European Wildlife and Natural Habitats, 1979) • Biogenetic Reserves under the Council of Europe • European Diploma Sites under the Council of Europe
B	• SSSIs and National Nature Reserves (NNRs) • Sites with limestone pavement orders (Wildlife and Countryside Act 1981 (WCA, 1981)) • Nature Conservation Review Sites (NCR) • Geological Conservation Review Sites • Marine Nature Reserves (WCA, 1981) • Areas of Special Protection for Birds (AOSPs) (WCA 1981) • Sites hosting Red Data Book (RDB) species • Sites hosting species in Schedules 1,5 and 8 WCA 1981
C	• Local Nature Reserves (LNRs) • Other sites (not described above) with BAP priority habitats/species • Sites of Importance to Nature Conservation (SINCs) and other local designations • Regionally Important Geological Sites (RIGs) • Other natural/semi-natural sites of significant biodiversity importance, not referred too above
D	Sites not in the above categories, but with some biodiversity or earth heritage interest
E	Sites with little or no biodiversity or earth heritage interest

Note: *Sites falling into more than one category should be classified into the most important category.*

For a thorough biodiversity evaluation it is essential that all habitats and sites are evaluated, not just those with designations, and that the evaluation criteria consider sites with BAP priority habitats and species. The NATA evaluation methodology, suggests that BAP priority habitats and species are incorporated in category C the 'County/high local conservation interest potential' category. As discussed above, to date there is no specific guidance on the weight to be given to BAPs in the planning process. However, as a general rule, it may be more appropriate to classify sites with national BAP priority species and/or habitats as category B sites (essentially of national importance) and sites with regional/local BAP species and/or habitats in category C, rather than classifying all BAP sites in category C. Obviously, as the guidance (DETR, 1998c) suggests it is important that some flexibility is applied in individual cases.

Use of the NATA evaluation categories to evaluate sites/habitats in road EISs would be useful in that it would provide a framework for standardising criteria that are used for evaluating nature conservation interests in different EISs, especially at a local level. This would enable comparisons between different schemes to be made more readily, and would overcome the inconsistency in the treatment of local sites identified in current practice. It would also ensure consistency between appraisal and EIA evaluation decisions.

The NATA guidance also sets out impact magnitude criteria and a set of decision rules for assessment of options on nature conservation, (which combine the evaluation and magnitude criteria). These could also be used at the project EIS stage and are explained in section 7. This level of detail will be particularly important for the evaluation of biodiversity value outside designated sites.

7. Impact prediction and assessment

After identifying the likely potential impacts on biodiversity during scoping and evaluating the importance of biodiversity receptors in the assessment of baseline conditions, the next stage is impact prediction and assessment. The following sections outline good practice methods of EIA impact prediction and provide guidance on assessing the significance of impacts on biodiversity.[18]

7.1 Methods of impact prediction

Methods of impact prediction

- *Direct measurements* eg of areas of habitat lost or affected, proportionate losses from species populations, habitats and communities.
- *Flowcharts and networks* can be used to identify chains of impacts and are therefore useful for identifying knock-on effects from direct impacts and classifying indirect impacts into secondary, tertiary, etc.
- *Quantitative predictive models* are useful as they can provide rigorously tested impact predictions as opposed to vague generalisations. However, the resource and time constraints for project EIAs often limit the use that can be made of models.
- *Geographical information systems (GIS)* are extremely useful for producing models of spatial relationships eg constraint maps (Treweek & Veitch, 1996).
- *Information from previous road projects* can be valuable, especially if impacts were quantified and monitored.
- *Expert opinion* is always needed for the interpretation of data. Where there is insufficient quantitative data, impact prediction has to rely on knowledge of potential impacts and biodiversity elements. Ideally, predictions based on expert opinion will be based on consultations of relevant experts.
- *Description and correlation* observed correlations between distribution and abundance of species and physical factors eg water regime, noise, can be used to predict the likely composition of biodiversity at a site where future physical conditions can be specified. For example, Dutch guidance on predicting the effects of motorway traffic on breeding birds (Reijnen *et al*, 1995) - Reference Box 5.
- *Experimental systems and field trials* can be used to quantify and validate ecosystem responses, but they can be costly and difficult to set up and will not always yield useful results within an EIA timescale
- *Habitat evaluation methods* eg US Army Corp of Engineers, 1990; US Fish and Wildlife Service, 1980 (discussed at section 6.3.1 above).

(English Nature, 1994b; Morris & Therivel, 1995; Treweek, 1999)

Box 25

Predictions in UK EISs often fall below the standards of the good practice guidelines (such as those summarised in the Box 27)[19]. For example, there are relatively few UK examples of quantitative impact prediction. Some other countries (eg the Netherlands and US) place much more emphasis on producing quantitative predictions and have issued guidance to facilitate this. Such guidance includes quantitative Habitat Evaluation Methods (see Table 4), a detailed manual on a range of prediction methods specifically for infrastructure projects (Road and Hydraulic Engineering Division, The

---

[18] These stages of EIAs are considered in more detail in English Nature (1994b), DoE (1995), Morris & Therivel (1995) and Treweek (1999) which are useful references - see Reference Box 5.
[19] Treweek *et al*, 1993; Thompson *et al*, 1997; Byron *et al*, 2000 - see Reference Box 5.

Netherlands Directorate-General for Public Works and Water Management, 1993 - see Reference Box 7) and Dutch guidance on predicting the effects of motorway traffic on breeding birds (Reijnen *et al*, 1995) noted in Box 25.

Quantitative predictions are generally considered to be preferable to qualitative (Duinker, 1987) as the former can be tested using monitoring (ie they can be regarded as hypotheses which can be tested using monitoring data (Buckley, 1991)) and appropriate triggers can be set to allow for modification of mitigation/compensation measures if the actual impacts are not as predicted. In reality many predictions are based largely on expert opinion and *'predictions may legitimately be based on any combination of speculation, professional judgement, experience, experimental evidence, quantitative modelling and other methods'* (Beanlands and Duinker, 1983) so long as the basis on which the predictions have been made is explicit (in Treweek, 1999). (See also Beanlands and Duinker (1984)). As impact predictions will generally involve a level of uncertainty, it is vital that the confidence/uncertainty in the predictions are discussed in the EIS (IAIA 1999 Biodiversity Working Group, unpublished). Ideally, an EIS should propose an explicit and consistent scale for expressing and ranking the uncertainty/confidence in the predictions. One example which could be used/adapted is shown below.

Table 9 - Level of confidence in predictions

Biodiversity Receptor				
Phase	Residual environmental effects significance rating for each impact	Level of confidence	Likelihood	
			Probability of occurrence	Scientific certainty
Construction	*Eg MAJOR, MODERATE, MINOR based on the significance criteria adopted for a particular EIA (see 7.2.2)*	*Eg LOW, MEDIUM, or HIGH*	*Probability of occurrence based on professional judgement Eg LOW, MEDIUM, or HIGH*	*Scientific certainty based on scientific information and statistical analysis or professional judgement Eg LOW, MEDIUM, or HIGH*
Operation				

(Adapted from Barnes and Davey, 1999)

Assessing the cumulative effects of the road project will include looking at the interactions among effects the project may cause on the environment, such as those between effects on water quality and on aquatic biodiversity resulting from sedimentation and run-off pollution. As with EIA in general there is no one approach or methodology for all assessments of cumulative effects. In some cases it may be possible to use modelling and geographic information systems. However, where information is lacking, qualitative approaches and best professional judgement should be used (CEAA, 1994 and 1999 - see Appendix 4). Examples of methodologies are discussed in more detail in CEAA (1999) and US CEQ (1997).

Highways Agency/DETR have commissioned work to produce guidance on strategic environmental assessment of road/transport programmes (a strategic equivalent to the current road EIA guidance manual Volume 11 DMRB (DoT, 1993)) which will include a cumulative effects assessment element. A consultation version of this guidance will be published later this year (P. Tomlinson, pers. comm.).

Questions to consider when analysing effects on biodiversity

(Based on CEAA, 1996a)

- What impact will the project have on the genetic composition of each species? Are different genotypes of the same species likely to be isolated from each other? To what extent will habitat or populations be fragmented?
- How will the proposal affect ecosystem processes? Is this proposal likely to make the ecosystem more vulnerable or susceptible to change?
- What abiotic effects will devolve – change in seasonal flows, temperature regime, soil loss, turbidity, nutrients, oxygen balance, etc?
- Is diversity measured at the species, community and ecosystem level?
- Is the biological resource in question at the limit of its range?
- Does the species demonstrate adaptability? Eg urban foxes habituate and adapt to street lighting
- Have sustainable yield calculations, including population dynamic parameters, been determined (eg lake capacities, population thresholds)?
- Is the data dependable? What are the sources used?
- Is the assessment based on long term ecological monitoring, baseline survey, reconnaissance level field observations and primary research?
- Are plans made throughout the assessment for meaningful data input from the public, non-government organisations and other stakeholders?
- What level of confidence or uncertainty can be assigned to interpretations of the effects?

Box 26

Good impact prediction/assessment practice is summarised in the following Box.

> **Summary of good impact prediction and assessment practice**
>
> - If possible, present the magnitude or physical extent of predicted impacts in quantifiable terms eg areas of land taken, percentage of habitat lost or numbers of communities, species or individuals affected. Place these in an international, national, regional or local context where appropriate.
> - Provide information on the nature of the impact, ie impact magnitude, duration, timing, probability, reversibility, potential for mitigation, likely success of mitigation, significance of impact before and after mitigation. It may be useful to summarise this information for each impact in a table. Information also needs to be provided on the cumulative effects of different impacts.
> - Seek to identify and address indirect impacts, which, in some cases, may be more important than direct impacts. They are however more difficult to predict and appropriate prediction methods should be used.
> - Describe the elements of wildlife and earth science interest affected, their importance, sensitivity, and ability to escape, relocate or adapt/habituate.
> - Describe impacts which may occur during construction and, if appropriate, decommissioning phases of the project as well as those arising during the operational phase.
> - Consider short or medium term as well as long term or permanent impacts; consider positive effects which might enhance nature conservation interest as well as negative effects.
> - Specify uncertainties in prediction.
> - Assess the significance of impacts likely to arise from the project against the projected baseline data rather than against existing conditions revealed in the field surveys. The EIS should describe the likely changes in biodiversity that would result without the project going ahead. For example, if the proposed project did not go ahead, traffic levels on the existing road may increase, leading to higher pollution levels with associated impacts on vegetation.
> - State the predicted post-mitigation significance of impacts ie the significance of residual impacts **after** all proposed mitigation measures have been taken into account.

Box 27

7.2 Assessment of impact significance

The key issues in impact prediction and assessment are identifying biodiversity elements likely to be affected and assessing the significance of impacts – either absolutely or by using a defined scale. It is essential that the criteria by which impact significance are judged are clearly set out in the EIS, though this is often not the case[20].

'The assessment should finish with a statement of the significance of the identified impacts, requiring interpretation of findings and valuing the conclusions. This process is necessarily subjective and should therefore be undertaken by an experienced ecologist.' (DoE, 1995). 'One of the most important parts of the EA process is to attach some measure of significance to impact predictions' (DoE, 1995).

This guidance suggests a systematic approach to assessing biodiversity impact significance should be followed to improve consistency between EIAs. One such approach could be based on the NATA methodology (DETR, 1998c) and this is

---

[20] Treweek *et al*, 1993; Thompson *et al*, 1997; Byron *et al*, 2000 - Reference Box 5.

discussed below. However, as this is not yet current practice (ie EISs generally do not use standard decision rules, but rather make assessments based on magnitude/significance criteria adopted on a one-off basis for a particular project) some examples of current good practice for determining impact magnitude and significance are also discussed.

7.2.1 NATA based approach

The NATA methodology gives generic impact magnitude categories (shown in Table 10). It is suggested that the principles set out in these definitions are used to formulate project specific magnitude criteria for each of the biodiversity receptors.

Table 10 - NATA Impact Magnitude Categories

Impact Magnitude category	Criteria
Major negative impact	'if, in light of full information, the proposal (either on its own or together with other proposals) may adversely affect the integrity of a site, in terms of the coherence of its ecological structure and function, across its whole area, that enables it to sustain the habitat, complex of habitats and /or the population levels of species for which it was classified'
Intermediate negative impact	'if, in light of full information, the site's integrity will not be adversely affected, but the effect on the site is likely to be significant in terms of its ecological objectives. If, even in the light of full information, it can not be clearly demonstrated that the proposal will not have an adverse effect on integrity, then the impact should be assessed as major negative'
Minor negative impact	'if neither of the above apply, but some minor negative impact is evident. In the case of Natura 2000 sites they may nevertheless require a further appropriate assessment if detailed plans are not yet available'
Positive impact	'where there is a net positive wildlife gain. Examples include a mitigation package where previously fragmented areas were united through habitat creation work (the concept of connectivity), a scheme which diverts traffic away from a designated site, and other proposals which do provide general wildlife gain through new design features such as hedges, ponds, ditches, scrub, linear woodland, grasslands and geological exposures. Many such improvements, while being very useful, will not provide a significant gain to the biodiversity interest within the Natural Area; these should be assessed as minor positive. However, where a significant net gain is evident, the features should be assessed as intermediate positive, or major positive if the net gain is one of national importance'
Neutral impact	'if none of the above apply, that is, no observable impact in either direction'

These criteria consider the impact of a proposal using the concepts of significance and integrity (NATA, 1998c). These criteria are reflected in the Habitats Directive and are applied in land use planning[21]. The concept of integrity is outlined in Box 28 and significance is discussed further in later sections of this guidance.

Red Squirrel

---

[21] DoE, 1994; SOED, 1995 and 1999; Welsh Office, 1996; DoE-Northern Ireland, 1997.

> ### The concept of 'integrity'
>
> *'The integrity of a site is the coherence of its ecological structure and function, across its whole area, that enables it to sustain the habitat, complex of habitats and/or levels of populations for which it was classified'* (Paragraph C10, PPG9 (DoE, 1994))
>
> In the Habitats Directive, this concept is used in relation to internationally designated sites (SACs and SPAs). However, this principle can be applied at all levels of sites in the conservation hierarchy and also to sites outside designated areas.

Box 28

Setting the criteria for what amounts to a 'high', 'medium' or 'low' magnitude impact for a particular project involves deciding what amount of change is acceptable in that case (sometimes referred to as the 'limits of acceptable change'). Ideally these criteria should be derived from appropriate objectives/targets for individual habitats and species. For example, for habitats subject to HAPs and species subject to SAPs, the targets in the appropriate action plans (national, regional and local) can be used to set the magnitude criteria, for natural areas/designated sites the conservation objectives/reasons why the site was initially designated can be used.

Some examples of magnitude criteria derived from national HAP/SAP and LBAP targets are given below.

Table 11 - Examples of criteria derived from National BAPs

BAP	Targets	Possible criteria
National Upland Oakwood HAP (HM Government, 1995b)	• Maintain the existing area of habitat and improve its condition • Expand the existing area of habitat by about 10% by some planting but particularly by natural regeneration by 2005 • Identify and encourage restoration of a similar area of former upland oakwood that has been degraded by planting with conifers or invasion by rhododendron	• Major negative – loss of any existing area of upland oak woodland • Moderate negative – predicted reduction in condition of an area of upland oak woodland • Minor negative – possible change in condition of an area of upland oak woodland • Beneficial – improved management of an area of existing upland oak woodland, restoration of an area of former upland oak woodland
National Natterjack toad SAP (HM Government, 1995b)	• Sustain all existing populations and where appropriate restore each population to its size in the 1970s	• Major negative – loss or fragmentation of existing habitat • Moderate negative – predicted reduction in condition of existing habitat • Minor negative – possible change in condition of existing habitat • Beneficial – improved management of an area of existing habitat, creation of addition new habitat

Table 12 - Examples of criteria derived from County BAPs

BAP	Targets	Possible criteria
Kent Chalk Grassland HAP (Kent Biodiversity Action Plan Steering Group, 1997)	• To ensure that all unimproved and semi-improved chalk grassland is under optimal management • To increase the extent of unimproved chalk grassland in the county • To create links between existing areas along the spine of the North Downs	• Major negative – loss of any existing area of unimproved chalk grassland in the county and/or severance/ fragmentation of links between existing areas of chalk grassland along the spine of the North Downs • Moderate negative – predicted reduction in condition of an area of unimproved chalk grassland in the county and/or predicted change in the management of an area of unimproved chalk grassland in the county which is likely to result in a reduction in condition • Minor negative – predicted reduction in condition of an area of semi-improved chalk grassland in the county and/or predicted change in the management of an area of semi-improved chalk grassland in the county which is likely to result in a reduction in condition • Beneficial – improved management of an area of existing unimproved and/or semi-improved chalk grassland, creation of additional new habitat, creation of links between existing areas of habitat
Kent Water Vole SAP (Kent Biodiversity Action Plan Steering Group, 1997)	• To arrest the decline in the water vole population in Kent by 2000	• Major negative – loss or fragmentation of existing water vole habitat and/or a decline in local populations • Moderate negative – predicted reduction in condition of existing water vole habitat and/or a decline in local populations • Minor negative – possible change in condition of existing water vole habitat • Beneficial – improved management of an area of existing water vole habitat, creation of addition new water vole habitat

Where there are no appropriate targets/nature conservation objectives, specific criteria will need to be developed on a case by case basis based on expert opinions/professional judgements. Ideally, the criteria will not be determined by the EIA consultant alone, but will involve the consultees (eg English Nature, Countryside Council for Wales, Scottish Natural Heritage, LBAP groups, RSPB, County Wildlife Trusts, local specialist groups, etc.).

The NATA methodology sets out decision rules to assist the assessment of options on nature conservation (DETR, 1998c – Annex 6B). These are presented in Table 13 in the form of a matrix showing the interaction between the nature conservation evaluation category (A to E) and the impact magnitude criteria (major negative to major positive).

Table 13 - NATA Decision Rules for Assessment of Options on Nature Conservation

Major +ve	Large positive	Large positive	Large positive	Large positive	Large positive
Intermediate +ve	Moderate positive	Moderate positive	Moderate positive	Moderate positive	Moderate positive
Minor +ve	Slight positive	Slight positive	Slight positive	Slight positive	Slight positive
Neutral	Neutral	Neutral	Neutral	Neutral	Neutral
Minor -ve	Slight adverse	Slight adverse	Slight adverse	Slight adverse	Neutral
Intermediate -ve	Large adverse	Large adverse	Moderate adverse	Slight adverse	Neutral
Major -ve	Very large adverse	Very large adverse	Large/ moderate adverse (see note 4)	Slight adverse	Neutral
	A	B	C	D	E

Notes:

1. Options that have a 'very large adverse effect' are likely to be unacceptable on nature conservation grounds alone (even with compensation proposals).
2. There should be a strong presumption against options in the 'large adverse' category, with more than 1:1 compensation (net gain within the Natural Area) for the very occasional cases where development is allowed as a last resort.
3. Options in the 'moderate adverse' category should include at least 1:1 compensation (no net loss within the Natural Area) if the development is allowed.
4. ...should score 'large adverse' if the habitats/species are not substitutable, or otherwise should score 'moderate adverse' (DETR, 1998c – Annex 6B).

The NATA guidance notes that applying the NATA decision rules becomes more complex when a project potentially affects more than one nature conservation feature and it suggests three rules which should be applied in these circumstances - see Table 14 (DETR, 1998c).

Table 14 - NATA Decision rules for multiple features

Rule	Explanation
Most damaging impact	'If a proposal affects, say five features, of which there is a 'large adverse' on one and 'slight adverse' on the other four, then the score should be 'large adverse'. The principle is that a proposal or option as a whole should be classified in the 'worst adverse' category if at least one site falls into this category. There may be a view that a scheme should not be marked down if only a single small feature is affected in this way. However, the rationale for this approach is that it encourages the development of alternative options which avoid such adverse outcomes.'
Cumulative adverse effects	'A proposal may affect a number of sites, each of which score 'slight adverse' or 'moderate adverse'. Where it is clear that the cumulative effect on all these sites is at least equivalent, ecologically, to a single site in a higher category, then the proposal should be scored in the higher category. Thus, for example, a proposal may affect 4 sites and have a moderate adverse assessment on each. If the view is that the cumulative effect is equivalent to a single site in the 'large adverse' category, then this score should be applied. It may be worth looking at examples across options or across proposals to help make this judgement appropriately and consistently.'
Positive effects	'When classifying a proposal or option with several sites, it may be appropriate to consider adverse assessments in some areas against a beneficial assessment (through mitigation, for example) elsewhere, to judge the net assessment overall. However, this assessment should not be based on a simple hectarage or number-of-sites approach; an appropriate ecological judgement has to be made about the overall effects of the proposal.'

Corncrake

Despite Note (1) to Table 13 and the *most damaging impact* principle above, in the recent roads review (DETR, 1998b) the A650 Bingley Relief Road scheme which was assessed as having a very large adverse effect on nature conservation was allowed to proceed (DETR, 1998d). To avoid these situations in the future, we recommend that a cut-off is agreed whereby any project scoring a very large adverse impact is automatically refused or alternatives are reconsidered.

7.2.2 Criteria used in recent EISs

Examples of magnitude and significance criteria used in recent EISs together with some guidance on assessing the magnitude of cumulative impacts on biodiversity are set out in Tables 15 - 17 below:

Table 15 - Magnitude criteria used in recent EISs

EIS	Magnitude Criteria
M25 Motorway Link Roads between Junctions 12 and 15 (DoT, 1994) A proposal to increase the capacity of the M25 between junction 12 (M3) to junction 15 (M4) by the addition of 2 or 3 lane link roads parallel and on either side of the existing M25 over a distance of approx. 7 miles. Also to widen the existing M25 through junctions 13, 14 and 15 and re-routing of part of the A30	Impacts are summarised as follows: • Major – loss of 5% or more of habitat or site. • Moderate – loss of up to 5 % of habitat or site, or predicted change in adjacent habitat. • Minor – no loss of habitat, or possible change in adjacent habitat.

Note: *These criteria are good in that they attempt to quantify magnitude through percentage of habitat affected. However, any criteria formulated using this approach should include a percentage of habitat or site loss that is ecologically relevant.*

Thames Water, Best Practicable Environmental Option Study (Thames Water, 1998) A strategic study on planning for future water resources	Impacts are classified as: • High – loss or damage to any site covered by a statutory (national) or international nature conservation designation eg SSSI, NNR, SPA. • Medium – loss or damage to a site covered by a local nature conservation designation. • Low – no loss or damage to sites covered by statutory designations or local nature conservation designations, but possible other damage eg to wetlands, hedgerows, woodland.

The magnitude of cumulative effects on biodiversity can be considered by first determining the separate effects of past, present actions, the proposed road project (and reasonable alternatives), and other future projects and activities. The cumulative effects on a specific receptor will not necessarily be the sum of all of the effects. To determine the cumulative effects it is essential to know how a particular receptor responds to environmental change. The assessment will need to consider whether effects will be additive, antagonistic or synergistic. It may be useful to summarise the cumulative effects in a table (US CEQ, 1997 - Appendix 4). For example:

Table 16 - Summary of cumulative effects

Resource	Past Actions	Present Actions	Proposed Actions	Future Actions	Cumulative Effect
Fish	Decrease in species numbers and diversity	Occasional documented fish kills	Increase in number of fish kills	Loss of cold-water species due to changes in temperature	Significant decline in numbers and species diversity
Wetlands	Large reduction in acreage of wetlands	Loss of small amount of wetland annually	Disturbance of a 5 acre wetland	Continued loss of wetlands	Significant cumulative loss of wetlands

(US CEQ, 1997)

Table 17 - Significance criteria used in recent EISs

EIS	Significance Criteria
A249 Iwade Bypass to Queenborough Improvement (Highways Agency, 1997) A 5.3 km off-line improvement to 2 lane dual carriageway with a bridge	'The severity of impacts would be judged on a number of characteristics that would include magnitude, spatial extent, duration and the nature/location of the impact. The significance of effects would be determined by combining the importance and sensitivity of the ecological resources…with the severity of impact. Categories of significance of effect are proposed as follows: *Major* • Permanent loss affecting the ability of the site to support internationally important habitat and the related species. • Adverse effect upon the integrity of the site, where the integrity of a site is the coherence of its ecological structure and function, across its whole area, that enables it to sustain the habitat, complex of habitats and/or levels of population of the species for which it was classified. • Permanent loss of any protected or nationally important rare species (as defined in Schedules 5 and 8 of the Wildlife and Countryside Act and the World Conservation Union Red Data Book (IUCN RDB) through loss of habitat, severance or disturbance. • Permanent loss of any priority habitat and species as defined under the EU Birds and Habitats Directive. • Permanent loss to those resources within a site of national importance where the presence of those resources were the reasons for the site's designation. *Moderate* • Permanent loss of nationally scare species (as defined in the relevant RDB) through loss of habitat, severance or disturbance. • Where an international or national site suffers some damage that compromises the ability of that site to support the habitats or species for which it was notified: but partial or total recovery is likely soon after cessation of the impact. • Where it only affects a small part of the site of national importance and to such a limited extent that the key elements of the ecosystem can continue to function. • Permanent loss of high quality of SNCI. *Minor* • Where a locally designated site suffers some damage that compromises the essential functioning of the habitat or species, but partial or total recovery is likely soon after the cessation of the impact. • Where it only affects a small part of the site of local importance and to such a limited extent that the key elements of the ecosystem can continue to function. Criteria for assessing the permanence of effects are also given as follows: • 'Permanent – Effects continuing beyond the span of one human generation (taken as over 25 years), which cannot be extinguished entirely. • Temporary – Where measures can be taken to reduce the effects over a length of time (under 25 years). • Long-term – 15 to 25 years or longer (eg: replacement of mature trees) • Medium-term – 5 to 15 years (eg: establishment of mature coppice). • Short-term – up to 5 years (eg: recreation of river habitats).'

A1 Motorway North of Leeming to Scotch Corner (Highways Agency, 1994) The improvement and conversion to motorway of a 10 mile section of the A1	Impact significance	Explanation	Impact
	EXTREME	Adverse impacts that are of international significance and thus represent key factors in the decision-making process. Typically no mitigation of the impact is possible. Effects may be such as to prevent a scheme from progressing.	*Any* impact on a site of international importance. *High* impact on a site of national importance.
	SEVERE	Adverse impacts that are of national significance and are important factors in the decision-making process. Mitigation of the adverse effects is not usually possible and if it is, there are likely to be residual impacts. Effects may be of such a scale as to radically influence scheme design.	Medium impact on a site of national importance.
	SUBSTANTIAL	Adverse impacts that are of county significance and are important factors in the decision-making process. Mitigation is usually possible to a certain extent but residual impacts are likely to remain. Will influence decision-making process but are not likely to be a deciding factor.	*Low* impact on a site of national importance. *Medium high* impact on a site of county importance.
	MODERATE	Adverse impacts that are of local significance and are likely to influence the decision-making process only if other factors are not an issue. The scope for mitigation is usually high, especially habitat creation.	*Low* impact on site of county importance. *Medium high* impact on a site of local importance.
	SLIGHT	Adverse impacts that are so small as to be of little or no significance.	*Low* impact on a site of local importance.

Blue Circle Medway Works EIS (Blue Circle, 1997)	Significance	Criteria
	SEVERE	Only adverse effects are assigned this level of importance as they represent key factors in the Town and Country Planning process. These effects are generally, but not exclusively associated with sites and features of national or regional importance. A change in a regional or district scale feature may also enter this category. Typically, mitigation measures are unlikely to remove such effects.
	MAJOR	These effects are likely to be important considerations at a local or district scale, but if adverse, are potential concerns to the project, depending upon the relative importance attached to the issue during the decision-making process. Mitigation measures and detailed design work are unlikely to remove all of the effects upon the affected communities or interests.
	MODERATE	These effects, if adverse, while important at a local scale, are not likely to be key decision-making issues. Nevertheless, the cumulative effect of such issues may lead to an increase in the overall effects on a particular area or on a particular resource. They represent issues where effects would be experienced but mitigation measures and detailed design work would ameliorate/enhance some of the consequences upon affected communities or interests. Some residual effects would still arise.
	MINOR	These effects may be raised as local issues but are unlikely to be of importance in the decision-making process. Nevertheless, they are of relevance in enhancing the subsequent design of the proposed development and consideration of mitigation or compensation measures.
	NONE	No effects or those which are beneath levels of perception, within normal bounds of variation or within the margin of forecasting error.

The key difference between determining the significance of direct/indirect impacts and determining the significance of cumulative effects is the influence of other projects and activities. The incremental cumulative effects of a particular project may be deemed to be significant when considered in the broader context of the effects of other projects and activities (CEAA, 1994 and 1999).

'The significance of effects...at the end of the day, usually relies on the professional judgement of the ecologist. This can therefore, lead to differences of opinion on the significance of impacts, because the ecologists may be placing varying weight on different factors.' (RSPB, 1995). Consequently it is essential that EISs clearly set out the reasoning behind assessments of impact magnitude and significance.

In current EIA practice, impact significance is generally determined by reference to the importance of biodiversity elements likely to be affected and the impact magnitude. However, impact *magnitude* is only one attribute of an impact and other attributes such as duration, timing, probability, etc. must also be taken into account. It is important therefore that these other attributes are described as fully as possible in the EIS and considered during the determination of impact significance.

Freshwater pearl mussel

8. Mitigation and enhancement

The key objective of this guidance not to significantly reduce biodiversity at any of its levels and to enhance biodiversity wherever possible and the guiding principle in Box 29 below should guide the design of mitigation and enhancement measures. As far as possible, all negative impacts should be mitigated not just those that are significant.

Guiding Principle

Avoid impacts on biodiversity and create opportunities for enhancement of biodiversity wherever possible by route selection and scheme design. Where this is not possible identify the best practical mitigation and enhancement option to ensure that there is no significant loss of biodiversity. Compensation should be viewed as a last resort.

Box 29

This guiding principle accords with the five-point approach to planning decisions for biodiversity (information, avoidance, mitigation, compensation and new benefits) that the RTPI (1999) propose. When an EIS discusses the significance of impacts this should be the significance after proposed mitigation measures have been taken into account, so it is clear what residual impacts will occur if the scheme proceeds. The EIS should give a precise description of the mitigation measures proposed, how these will be implemented, their status (ie whether the developer has given a firm undertaking to carry out measures or whether they are 'recommendations') and a clear assessment of likely success of the proposed mitigation/enhancement measures. Treatment of these issues in EISs is often poor (DETR, 1997; Byron *et al*, 2000). An example of criteria for assessing the effectiveness of mitigation measures is given below.

Table 18 - Criteria for assessing effectiveness of mitigation

A322 Improvement – Bisley Common to Brookwood Crossroads (Surrey County Council, 1995)	'The effectiveness of recommended mitigation measures are evaluated on the following basis: • Poor – some mitigation but little overall reduction in impact. • Limited – the mitigation measures reduce the impact to some degree. • Moderate – reasonable mitigation, but original impact will still be felt to a significant degree. • Substantial – almost complete mitigation.'

EISs should distinguish between the types of mitigating measures proposed ie whether they are avoidance, mitigation, compensatory or genuine enhancement measures. Worryingly, RSPB (1995) found that some EIAs were falsely claiming 'enhancement' to gain advantage in the decision-making process and the terms mitigation and compensation are often incorrectly used synonymously (D. Hill, Personal communication). To avoid confusion, the EIS should define how it is using these terms eg see the definitions in the following box.

> ### Definitions of avoidance, mitigation, compensation and enhancement
>
> (Based on DoE, 1995; RSPB, 1995)
>
> **Avoidance**
>
> Measures taken to avoid adverse impacts, such as locating the main development and its working areas and access routes away from areas of high ecological interest, fencing off sensitive areas during the construction period, or timing works to avoid sensitive periods. Also includes alternative or 'do nothing' options.
>
> **Mitigation**
>
> Measures taken to reduce adverse impacts eg modifications or additions to the design of the development, such as the creation of reed bed silt traps to prevent polluted water running directly into ecologically important watercourses. The preservation of 'wildlife corridors' between habitats which would be separated by a proposed development may reduce the possible effects on some fauna.
>
> **Compensation**
>
> Measures taken to offset/compensate for residual adverse effects which cannot be entirely mitigated. These usually take the form of replacing (or at least trying to) what will be lost eg the relocation of important grassland or heathland habitats from the development site to another area identified as suitable (using techniques such as soil or turf transfer), or the creation of new habitats.
>
> **Enhancement**
>
> The genuine enhancement of biodiversity interest eg improved management or new habitats or features, with the result that there is a **new** benefit to biodiversity ie improvements over and above those required for mitigation/compensation.

Box 30

Road EISs often fail to consider the full range of possible avoidance, mitigation, compensation measures (Treweek *et al*, 1993; Byron *et al*, 2000). The checklist of possible options in Box 31 should help parties involved in the EIA process to identify measures which may be appropriate for particular schemes.

Checklist of potential avoidance, mitigation, compensation and enhancement measures

Avoidance
- Route alignment to avoid loss and/or severance of sensitive areas or disturbance during construction

Mitigation
- Use of bridges and viaducts where embankments may change water levels leading to adverse effects on wetlands
- Careful drainage design eg use of balancing ponds to reduce pollution/provide additional capacity to cope with stormwater/provide valuable habitats in own right
- Planting; native species, replace hedgerows to maintain or increase connectivity
- Ecopassages
- Specialised fencing
- Specialised lighting eg specialised lighting used for nightjars on the road across Chobham Common (D. Hill, Personal communication)
- Landform: irregularities of slope of cuttings/embankments to provide greater range of microhabitats
- Re-establishing ecotones and buffers
- Allow natural regeneration (ie no seeding/planting) where appropriate
- Water quality mitigation techniques which will prevent/reduce impacts on aquatic biodiversity include:
    i) grit/silt traps
    ii) oil interceptors
    iii) french drains
    iv) sedimentation tanks/lagoons
    v) grass swales
    vi) aquatic/vegetative systems
    vii) pollution traps
    viii) straightening

- Flood plain mitigation measures - soft engineering solutions working with natural systems (eg tree-lined banks, wetland shallows) should be used wherever possible. Creation of artificial watercourses eg culverts should only be used as a last resort. Other measures to consider include:
    i) flood plain improvements, including removal of existing obstructions to flow, compensatory flood storage and new openings in existing embankments
    ii) local flood protection measures eg flood walls and flood protection embankments
    iii) improvements to existing river structures
    iv) channel improvements, including deepening, widening, draining and straightening

(cont...)

Box 31

> ### Checklist of potential avoidance, mitigation, compensation and enhancement measures (cont)
>
> **Compensation**
> - Translocation/re-establishment of habitats
> - Habitat restoration
> - Habitat creation – ideally 'in kind' ie creating habitat of the same type and quality as that which has been lost
> - Management plans for particular sites especially where habitats to be created/restored
> - Mitigation banking ie obtaining and restoring/creating and/or managing compensation sites to be used as credits against which habitat losses from a particular project can be 'traded'. To date most commonly used in the US in relation to wetlands
>
> **Enhancement**
> - Planting; native species, replace hedgerows to maintain or increase connectivity
> - Ecopassages
> - Establishing/re-establishing ecotones and buffers
> - Management plans for particular sites especially where habitats to be created

Box 31 cont

Useful references which discuss specific avoidance, mitigation, compensation, and enhancement measures in more detail are included in Reference Boxes 7 and 8.

There appears to be an emphasis on habitat creation and translocation in recent EISs. While this may be acceptable as a measure of last resort in some particular cases it **must not** be used as a justification to allow adverse impacts on high value biodiversity receptors (see Box 32).

Habitat creation and translocation

(English Nature, 1994a; Gault, 1997; RTPI, 1999 - See Reference Boxes 7 & 8)

- 'In the case studies examined, habitat creation and translocation were frequently encountered as mitigation for damage to SSSIs and other important sites. However, from the research available, it has to be concluded that these measures are **totally unacceptable** as mitigation unless it can be shown that the site can be re-created in full at minimum risk, and within a short time span.....' (English Nature, 1994a)

- 'In most cases the high value sites consist of long-established habitats of great complexity, with small scale variation in plant and animal communities reflecting the underlying patterns of soils and ambient environmental factors, and the reasons for the complex, inter-related patterns found are not fully understood. It is impossible, therefore, to re-establish them' (English Nature, 1994a)

- 'Habitat translocation has been attempted in many situations to rescue something of the threatened habitats. In many respects this **can** (if carried out proficiently) re-create a better resemblance to the original habitat than habitat creation because it is re-using soils and a proportion of the plant life. In some cases, some animals may also be transferred. Habitat translocation can be regarded as the best way of re-using material that is worth keeping, but which is not derived from a high value habitat. The dividing line between the acceptable and unacceptable use of habitat translocation for nature conservation is a fine one. It can be used for scheme enhancement, as a building block for habitat creation, but **it does not provide compensation for loss or damage to high value, non-replaceable sites...**' (English Nature, 1994a)

- **'It must be concluded that neither habitat creation nor translocation provide compensation or acceptable mitigation for the loss of all or part of high value sites'** (English Nature, 1994a)

- 'Where an irreplaceable site faces destruction, translocation may be the best form of mitigation. It is not the only form of mitigation and should only be considered in the context of other options such as purchase and suitable management of land and enhancement of similar but lower quality sites.' (Gault, 1997)

- Where development is proposed on a site of nature conservation interest, which would harm that interest, translocation may be offered as a form of mitigation. However, the real chances of success are usually low, even for those species and habitats that may be relocated; many cannot be moved. Translocation is not a substitute for in situ conservation. It cannot avoid demonstrable harm, or compensate for the loss of nature conservation value and it cannot remove the proposal's conflict with policies intended to protect habitats and species. It should not be taken into account until the planning decision has been made, weighting the benefits of the developments against the harm to biodiversity conservation. If the development proceeds, despite the harm, then translocation is essentially a rescue operation where nothing would be lost by trying to move the species or habitat' (RPTI, 1999).

Meadow

Box 32

The EIS should set out how the mitigation measures will be implemented. It is important that the EIA integrates biodiversity mitigation measures with other mitigation measures (eg landscape, water, cultural, etc.) to avoid conflicts between the objectives of the different mitigation measures.

> **Key questions to ask about proposed mitigation measures**
> (World Bank, 1997)
>
> - Does the project address issues concerning the integrity of natural habitats and ecosystems and maintenance of their functions?
> - Do the project boundaries encompass the relevant natural habitats/ecosystems within limitations of political and administrative boundaries? Have adequate steps been taken to deal with issues affecting ecosystems outside the project boundaries?
> - Have local communities dependent on the affected area(s) been included in the preparation and implementation of the project? Are arrangements agreed on compensation and/or concessions to groups adversely affected by the project?
> - Is the project design flexible enough to manage the predicted changes? Does the project draw adequately upon scientific and local knowledge to inform adaptive management of the natural environment?
> - Does the project involve all the relevant sectors and disciplines?

Box 33

As mitigation measures discussed in EISs are not binding, planning authorities should ensure that detailed prescriptions are incorporated into planning conditions or obligations to ensure that the implementation of these measures is enforceable. Alternatively, an Environmental Action Plan or Environmental Management Plan can be used to operationalise the mitigation measures (see section 11 below). A summary checklist to help ensure effective mitigation is included in Box 34.

> **Effective mitigation checklist**
>
> in this box the term mitigation is used to mean all avoidance, mitigation, compensation and enhancement measures proposed for a particular project
>
> - Consider mitigation at the outset of the project
> - The mitigation measures proposed must be feasible with defined criteria for success eg retaining all or part of a target proportion of an extant population or habitat
> - It is important to be confident that the proposed mitigation measures can be achieved eg the translocation of existing or creation of new habitat may depend on certain soil conditions, so it is important at the outset to be confident that these can be achieved on site.
> - Consider the importance of ecological processes (eg the population dynamics) in relation to the proposed mitigation measures. For example if like-for-like habitat is being provided for an affected species it is vital that the powers of dispersal/colonisation of the particular species are considered. Where an EIS claims that displaced individuals can move to 'available' habitat in the vicinity, the EIS must contain an assessment of whether that habitat is really 'available' ie is it already at capacity/sub-optimal?
> - Assess the impacts of proposed mitigation measures
> - Secure adequate funding
> - Prepare an Environmental Management Plan/Environmental Action Plan/Conservation Management Plan to provide for implementation and monitoring of mitigation measures (see section 11 below)
>
> (D. Hill, Personal communication)

Box 34

There is a need for mitigation to be placed in the wider framework of cumulative effects and biodiversity maintenance (this could be achieved via strategic environmental assessment (SEA), sustainability appraisal, or cumulative effects assessment) (IAIA 1999 Biodiversity Working Group, unpublished). The Transport Research Laboratory (TRL) is currently carrying out work on a possible methodology for assessing the cumulative effects of the UK road network. Without this wider perspective there is a great danger that project EIAs, no matter how systematic, will collectively contribute to the incremental reduction of the 'UK biodiversity baseline'.

9. Presentation of biodiversity information in EISs

The raison d'être for an EIA is to inform the decision-making process. Therefore, it is imperative that the EIS be comprehensible to decision-makers. Thus it should be concise, informative and succinct. Box 35 below sets out general EIS preparation advice. The parties preparing the EIS should refer to the EIA Directive (Article 5 and Annex IV) requirements as to the information that should be included in an EIS – see Box 16 on page 29.

General EIS preparation advice

(DoT, 1993 Section 4 part 3)

The EIS should be in three parts:

Volume one - a comprehensive and concise document drawing together all the relevant information about the scheme.

Volume two - a volume containing a detailed assessment of significant effects by subject area. This will not be necessary where there are no significant effects.

Non-technical summary (NTS) - a brief report summarising the principle sections of volume one of the EIS in non-technical language which is readily understandable by members of the public.

Proposed method

The following headings are suggested for organising the information required in volume one of the EIS:

- Introduction
- The Existing Traffic or Environmental Problem and the Proposed Scheme
- The Proposed Scheme
- Baseline Information
- Mitigation
- Environmental Effects
- Route Options
- Consultations
- Environmental Impact Tables (EITs). The EIT is a tabular presentation of data summarising the main likely direct and indirect impacts of the proposed highway scheme taking account of agreed mitigation. There should be a specific part of the EIT summarising the impacts on the Cultural and Natural Environment and within this a summary of ecological impacts of the preferred route compared with a 'do-nothing' or 'do-minimum' option.

Box 35

One good format for organising the environmental effects section of an EIS is to set out each of the issues raised at the scoping stage/in the scoping report, then to discuss these in turn explaining how each has been addressed and the degree of confidence in the predictions and mitigation proposals. It is useful for those reviewing the EIS as part of the decision making process if the terms of reference of each of the specialist studies are also included in the EIS.

Biodiversity information in the EIS - summary of good practice

- Include a 'biodiversity method statement' describing:
  - the specialist ecologist company/individuals responsible for the biodiversity part of the EIS and terms of reference (TOR) for specialist studies
  - the scoping process including planning new surveys and the areas considered but not dealt with in detail and the reasons for this.
  - the level of contact with biodiversity consultees
  - criteria used to evaluate: the importance of biodiversity elements, the magnitude of impacts, the significance of impacts, the likely success of proposed mitigation/enhancement measures
  - any guidelines, methods or techniques used.

- Include clear colour coded or annotated maps, showing:
  - the study areas considered
  - biodiversity constraints including designated areas and areas subject to BAPs/LBAPs
  - the different types and quality of all habitats likely to be affected.

- An assessment of the biodiversity impacts of the alternatives considered.
- Reference all sources of background information eg research papers and existing data.
- Include or clearly reference all new data collected for the EIS. (generally put data in appendices to limit the size of the text of the actual EIS.) State collection methods, survey timing and duration, and limitations.
- The length and detail of the descriptions of effects should reflect their relative importance.
- Give as full a **factual** description as possible of predicted impacts: impacts should be quantified as far as is practicable; any judgements made on the advice of statutory or other expert consultees should be noted. The aim is to provide sufficient data to allow decision-makers to form their own judgements about the significance of impacts.
- Cumulative effects on biodiversity can be discussed either in a separate section or as an integral part of the analysis of biodiversity impacts.
- Explain the proposed mitigation and enhancement measures, give detailed prescriptions for their implementation and assess their likely success.
- Summarise the residual impacts on biodiversity after mitigation.
- Describe the proposed biodiversity post-project monitoring programme: what will be measured, when, how, by whom.
- Explain how and by whom unexpected post-project impacts will be remedied.

Box 36

None of the EISs reviewed in detail (Byron *et al*, 2000) specifically discussed impacts on biodiversity, but EISs which appeared to have considered most ecological issues and may therefore be useful as examples of current practice and presentation include M25 Motorway Link Roads between junctions 12 and 15 (DoT, 1994), A1 Motorway North of Leeming to Scotch Corner (Highways Agency, 1994), and A429 Barford Bypass (Warwickshire County Council, 1996).

10. Decision-making

Pursuant to the EIA Directive (as amended) and the UK implementing EIA Regulations (see Table 2) the competent authority cannot make a decision on a project until it has taken into consideration the EIS (including any further information requested by the competent authority) and any representations about the environmental effects of the development made by a member of the public likely to be affected or any of the consultation bodies.

The competent authority must publish its decision together with a statement that it has complied with these requirements. This statement should also contain:

- The contents of the decision and any conditions attached to it;
- The main reasons and considerations on which the decision is based, including the reasons for the option chosen and why any alternatives were rejected; and
- Where the decision is to proceed with the construction or improvement, a description of the main measures to avoid, reduce, and if possible, offset the major adverse effects of the project.

In the UK, (unlike some other countries eg the Netherlands) there is no formal review/audit of the detailed reports (including the biodiversity/ecology reports) prepared for the EIA by officially appointed bodies.

Woodlark

11. Biodiversity monitoring programmes and environmental management plans

Monitoring is not required by the EIA Directive. Indeed post-project monitoring is probably the weakest area of current EIA practice. However, inclusion of monitoring programmes is vital to provide a 'feed-back loop' enabling evaluation of the predictions of the EIS, the success of mitigation measures to be judged and post-development problems to be identified and rectified. As well as these 'project-specific' benefits, monitoring can also provide valuable information for use in future EIAs and for improving the science base of EIAs generally.

The need for monitoring

'Monitoring methods should be established in prediction and mitigation stages of the study and biodiversity data obtained through monitoring should be included in global data services such as the CHM [Clearing House Mechanism] and BCIS [Biodiversity Conservation Information System]' (Bagri *et al*, 1998)

'Monitoring is essential to understanding the effects of a project and to evaluating the degree of implementation and the success or failure of mitigation efforts (CEAA, 1996a)

'Where the success of mitigation is unclear or where failure might lead to very significant effects, it may be important to monitor mitigation measures, so that they can be corrected or redesigned if they are not sufficiently effective.' (English Nature, 1994b).

Box 37

However, despite the potential benefits, the vast majority of current UK road EIAs do not include a commitment to monitoring. In the review by Byron *et al* (2000) only 5% of EISs included a commitment to monitoring some aspect of the scheme, monitoring as a possibility for the future was discussed in 10% of the EISs.

Ecological monitoring involves the systematic observation and measurement of ecosystems (or their components) to establish their characteristics and changes over time (Treweek, 1999). Spellerberg (1991) discusses the principles of ecological monitoring in detail. It is important that the monitoring programme is well structured. Ideally the monitoring programme will include monitoring at each of the project stages (ie pre-construction and during construction as well as once a road is in operation). It is crucial that standard techniques/methods of data collection are used (and made explicit) so that the data can be used for comparative purposes. A good monitoring programme should be structured to address clearly defined questions, it will provide for repeatability and control and will have established appropriate timing and frequency in relation to the biodiversity elements being measured and the nature of the intended/implemented road project. There needs to be a quality control mechanism for assessing the monitoring data that should be independent to have credibility. For example a Conservation/Monitoring Group of interested parties such as the planning authority, the developer, the consultants, consultees, etc. could be set up. Box 38 summarises the key elements in developing a monitoring programme.

> **Developing a monitoring programme**
>
> Many of the elements necessary for adequate monitoring will have been developed as part of project planning and environmental analysis. This include the following:
>
> - Gathering data
> - Establishing baseline conditions
> - Identifying ecological elements at risk
> - Selecting ecological goals and objectives
> - Predicting the likely project impacts
> - Establishing the objectives of mitigation.
>
> The following additional monitoring-specific steps can build upon these elements:
>
> - Formulate specific questions to be answered by monitoring
> - Select indicators
> - Identify control areas/treatments
> - Design and implement monitoring
> - Confirm relationship between indicators and goals and objectives
> - Analyse trends and recommend changes to management.
>
> The breadth and specificity of the monitoring program will be determined by the biodiversity goals and objectives established as part of project planning and environmental analysis. (US CEQ, 1993)

Box 38

Manchester University EIA Centre (1999) and Treweek (1999) discuss the role of monitoring and post-auditing in the EIA process in more detail. It is essential that both the biodiversity data collected for the EIS and any subsequent monitoring data are made as widely available as possible eg to local communities, appropriate authorities, and biodiversity information networks. UK data should be fed in to the UK National Biodiversity Network. Ideally, monitoring associated with an EIA could contribute to wider/longer term monitoring programmes such as monitoring of priority habitats/species pursuant to HAP/SAP objectives/targets.

Environmental Management Plans (EMPs) (also referred to as Environmental Action Plans and Conservation Management Plans), although not required by UK EIA legislation can be used to operationalise proposed EIA mitigation measures and monitoring procedures (T. Dorken, Personal communication; Environment Agency, 1998, 1999; World Bank Environment Department, 1999). For example, if the proposed A465 Abergavenny to Hirwaun Dualling scheme proceeds an EMP will be prepared (Welsh Office Highways Directorate, 1997).

Such plans can provide a framework for implementation of mitigation measures, carrying out monitoring and on-going management of a road. They need to include: prescriptions, a work programme, schedules, be for an appropriate timescale (eg new habitats will require long term management), targets, a monitoring programme, a quality control mechanism for reviewing the monitoring data, and provisions for remedial action if the mitigation/management targets are not achieved. Environment Agency publications (1998, 1999) discuss suggested structures/content of plans in further detail.

Part III – Review

This section provides a checklist of key questions to ensure that the Guiding Principles have been implemented. Ideally the answer to each of the key questions should be 'yes', where this is not the case the issue should be reconsidered.

Guiding Principles	Key Questions
✓ Avoid impacts on biodiversity and create opportunities for enhancement of biodiversity wherever possible by route selection and scheme design. Where this is not possible identify the best practical mitigation and enhancement option to ensure that there is no significant loss of biodiversity. Compensation measures such as translocation should be viewed as a last resort.	• Have all impacts on biodiversity been avoided wherever possible? • Have all unavoidable impacts on biodiversity been reduced as far as possible? • Does the scheme ensure that there is not a significant loss of biodiversity? • If the scheme involves compensation eg habitat creation is this likely to be successful? • Have opportunities for enhancement been considered?
✓ Apply the precautionary principle to avoid irreversible losses of biodiversity. ie where an activity raises threats or harm to biodiversity precautionary measures should be taken even if certain cause and effect relationships are not scientifically established.	• In each case where it is not possible to thoroughly assess but it is suspected that there maybe an impact on biodiversity have avoidance/mitigation measures been incorporated?
✓ Widen existing EIA practice to an ecosystem perspective – ie consider the impacts of a road scheme on biodiversity and possible enhancements of biodiversity in the context of local and regional ecosystems, not just the immediate vicinity of the road.	• Has the EIA considered impacts and enhancements of biodiversity in the context of local and regional ecosystems?
✓ Safeguard genetic resources by protecting the higher levels of biodiversity (ie individuals, populations, species and communities, etc.) and the environmental processes which sustain them.	• Has the EIA considered all of the levels of biodiversity? • Has the EIA considered environmental processes?
✓ Consider the full range of impacts on biodiversity eg indirect and cumulative impacts not just the direct impacts such as species and habitat loss.	• Has the EIA considered direct impacts on biodiversity? • Has the EIA considered indirect impacts on biodiversity? • Has the EIA considered cumulative impacts on biodiversity? • Has each impact been discussed in detail and quantified wherever possible?

Guiding Principles	Key Questions
✓ The study area of the scheme should reflect the impact type (eg indirect effects will often extend throughout a watershed) rather than taking a fixed width corridor approach.	• Does the EIS clearly explain the study area chosen and the rationale for selecting this? • Does the EIS discuss the study area for each impact type? • Is the study area for the EIA appropriate for the consideration of cumulative impacts?

✓ Evaluate the impacts of a road scheme on biodiversity in local, regional, national, and, where relevant, international contexts ie an impact could be minor locally but significant at a national level eg where the locality has a very high proportion of a national rare biodiversity resource.	• Does the EIS clearly set out the following and explain why these where chosen:
	  i) criteria for determining receptor importance?
	  ii) criteria for determining the magnitude of each impact type?
	  iii) criteria for determining the significance of each impact type?
	• Has the significance of each impact after mitigation been considered at each of local, regional, national, and where relevant international, levels?

✓ Retain the existing pattern and connectivity of habitats eg protect natural corridors and migration routes and avoid artificial barriers. Where existing habitat is fragmented implement measures eg tunnels, bridges to enhance connectivity.	• Does the scheme preserve the existing habitat connectivity by route selection and/or the inclusion of animal tunnels and bridges? • Have connectivity enhancement measures been included where appropriate? • Have full details of the design and installation of each tunnel and bridge been set out in the EIS?
✓ Use buffers to protect important biodiversity areas wherever possible	• Have buffers been used to protect important biodiversity areas wherever possible?
✓ Maintain natural ecosystem processes in particular hydrology and water quality. Wherever possible use soft engineering solutions to minimise impacts on hydrology.	• Have impacts on ecosystem processes been avoided or minimised as far as possible? • Have soft engineering solutions been incorporated where appropriate?
✓ Strive to maintain/enhance natural structural and functional diversity eg ensure that the quality of habitats and communities is not diminished and wherever possible is enhanced by the road scheme.	• Does the scheme preserve the quality of each habitat/community? • Where a habitat/community is not currently of high quality have enhancement measures to improve habitat/community quality been incorporated?
✓ Maintain/enhance rare and ecologically important species (key species) - ie protected species SAP species, characteristic species for each habitat as loss of these may affect a large number of other species and can affect overall ecosystem structure and function.	• Does the EIS clearly set out the key species considered and explain why each of these has been chosen? • Have impacts on key species been avoided wherever possible? • Have unavoidable impacts on key species been mitigated as far as possible? • Have measures to enhance the status of key species been included?

Guiding Principles	Key Questions
✓ Decisions on biodiversity should be based on full information and monitoring must be part of the EIA process. The results of monitoring should be available to allow evaluation of the accuracy of impact prediction and should be widely circulated to help improve future road scheme design and mitigation.	• Have all appropriate background sources of biodiversity data been utilised? • Has relevant scientific literature been consulted? • Have all necessary surveys been carried out? • Do habitat/community surveys include an assessment of the quality of each habitat/community? • Has abundance and distribution data been collected for each key species? • Are the results of each survey included or referenced in the EIS? • For each survey does the EIS record: the date, the duration, the methodology, the qualifications of the person/people who carried out the work? • Has a monitoring programme been devised and explained in the EIS? • Will funding be made available to ensure that the monitoring programme goes ahead? • Will a quality control mechanism for reviewing the monitoring data (eg a Conservation/Monitoring Group of interested parties) be put in place? • Will the results of the monitoring programme be disseminated as widely as possible?
✓ Implement ongoing monitoring/management plans for existing and newly created habitats and other mitigation, compensation and enhancement measures.	• Will an ongoing monitoring/management plan be devised? • Has an outline of a proposed plan been described in the EIS? • Does the plan provide for remedial action if mitigation/management targets are not achieved?

Use of this guidance should help ensure that the potential impacts on biodiversity are thoroughly and explicitly addressed in road EIAs and that these EIAs interface more closely with the UK biodiversity process and the available research literature. Like any guidance undoubtedly this guidance will evolve through use, but it aims to provide a starting point for systematic assessments of biodiversity in road EIAs.

Acknowledgements

Many thanks are due to all of the following people who agreed to be interviewed and/or reviewed an early draft of this guidance and gave their valuable time and advice:

Ruth Adams	Cornwall Wildlife Trust
Penny Anderson	Penny Anderson Associates
Penny Angold*	Birmingham University
Andrea Bagri	IUCN, Gland
Olivia Bina	ERM
Ruud Cuperus*	Road and Hydraulic Engineering Division, Ministry of Transport and Public Works, The Netherlands
Tim Dorken	Highways Directorate, Welsh National Assembly
Jeff Edwards	Hampshire County Council
John Edwards	Surrey County Council
Ralph Gaines	London Wildlife Trust
Charlotte Gault	then at Cornwall Wildlife Trust
Richard Graves	London Borough of Bromley
David Hill	Ecoscope
David le Maitre	Environmentek CSIR, South Africa
William Latimer	WS Atkins
John Lawton	then at Imperial College London
Caroline Lidgett	Warwickshire Field Services
Alan Moreton	Imperial College, London
Peter Oggier*	University of Berne, Switzerland
Mike Oxford	North Somerset District Council
John Prendergast	then at Imperial College, London
Jost Rotar*	Ministry of Finance, Republic of Slovenia
Terry Rowell	Countryside Council for Wales
Sally Russell	Imperial College, London
Barry Sadler	IEAM
Tony Sangwine*	Highways Agency
Martin Slater	Environment Agency
Ian Spellerberg	Lincoln University, New Zealand
Jo Taylor	Nottingham Biodiversity Action Group
Stewart Thompson	Oxford Brookes University
Paul Tomlinson	TRL
Jo Treweek	Komex Clarke Bond
David Tyldesley	David Tyldesley Associates
Len Wyatt	Highways Agency

*Infra Eco Network Europe (IENE) co-ordinators

Special thanks are due to Bill Sheate at Imperial College, London, and to past and present members of the Transport and Biodiversity Group:

Jill Barton	Surrey Wildlife Trust
Barnaby Briggs	RSPB, then ERM, now Shell
Clare Brooke	RSPB, now Environment Agency
Carol Hatton	WWF UK
Rowena Langston	RSPB
David Markham	English Nature

Abbreviations

BAPs	Biodiversity Action Plans
CBD	The 1992 Convention on Biological Diversity
CHM	Clearing House Mechanism
DETR	Department of the Environment, Transport & the Regions
DoE	Department of the Environment
EA	The same as EIA
EIA	Environmental Impact Assessment
EIA Amendment Directive	EC Directive 97/11
EIA Directive	EC Directive 85/337 on Environmental Impact Assessment
EIS	Environmental Impact Statement
ES	The same as EIS
HAPs	Habitat Action Plans
HSs	Habitat Statements
LBAPs	Local Biodiversity Action Plans
NATA	DETR's New Approach To Appraisal of trunk road schemes
SAPs	Species Action Plans
SBSTTA	The CBD's Subsidiary Body for Scientific, Technical, and Technological Advice
UK BAP	The UK Biodiversity Action Plan

Glossary

Abiotic	Not biotic, not of life. Part of the environment which is not biological; that is water, soil, climate, geology (Spellerberg and Sawyer, 1999)
Allele	Different forms of a particular gene
Assemblage	A group of species characteristically found in the same location due to the similarity of their habitat requirements (English Nature, 1998d)
Avoidance	Measures taken to avoid adverse impacts, such as locating the main development and its working areas and access routes away from areas of high ecological interest, fencing off sensitive areas during the construction period, or timing works to avoid sensitive periods. Also includes alternative and 'do nothing' options
Baseline conditions	The conditions that would pertain in the absence of the proposed action
Biodiversity	The total range of variability among systems and organisms at the following levels of organisation: bioregional, landscape, ecosystem, habitat, communities, species, populations, individuals, genes and the structural and functional relationships within and between these different levels
Bioregion	Coherent natural area defined by landscape and species. Contrasts to regions defined by artificial political boundaries (Jeffries, 1997)
Biogeographic	Pertaining to the geographical distribution of living organisms, past and present, their habitats and their ecological interrelationships (English Nature, 1998d)
Biotic	Pertaining to living organisms or life
Biotic community	Populations of different species living together (Spellerberg and Sawyer, 1999)
Boundary	A zone composed of the edges of adjacent ecosystems (Forman, 1995)
Buffer zone	An area or zone that helps to protect a habitat from damage, disturbance or pollution. It is an area (human-made or natural) that is managed to protect the 'integrity' of that area (Spellerberg and Sawyer, 1999)
Carrying capacity	The maximum number of organisms or amount of biomass that can be supported in a given area (Treweek, 1999)
Compensation	Measures taken to offset/compensate for residual adverse effects which cannot be entirely mitigated. These usually take the form of replacing (or at least trying to) what will be lost eg the relocation of important grassland or heathland habitats from the development site to another area identified as suitable (using techniques such as soil or turf transfer), or the creation of new habitats
Competent authority	The authority which determines whether or not an application for a project can proceed

Sand lizard

Configuration	A specific arrangement of spatial elements that is found in different places (Forman, 1995)
Connectivity	A measure of how connected or spatially continuous a corridor, network, or matrix is. (The fewer the gaps the higher the connectivity). Related to the structural connectivity concept; functional or behavioural connectivity refers to how connected an area is for a process, such as an animal moving through different types of landscape elements (Forman, 1995)
Corridor	A strip of a particular type that differs from the adjacent land on both sides. (Corridors have several important functions, including conduit, barrier, and habitat) (Forman, 1995)
Cumulative environmental effects	Effects on the environment that are caused by a project in combination with effects those of other past, present and future projects and activities
Direct impact	An outcome that is directly attributable to a defined action (Treweek, 1999)
Dispersal	The spreading of an organism's propogules (eg seeds, spores) (Spellerberg and Sawyer, 1999)
Dispersion	The spatial pattern of distribution of organisms (Spellerberg and Sawyer, 1999)
Disturbance	Disruption of normal process or behaviour
Ecology	The science of the interrelationships between living organisms and their environment (other organisms and the physical environment including the soil, air, climate) (Spellerberg and Sawyer, 1999)
Ecological impact assessment	The process of defining, quantifying and evaluating the potential impacts of assessment defined actions on ecosystems or their components (Treweek, 1999)
Ecopassages	All sorts of tunnels/underpasses/ecoducts/'green bridges' by which wildlife can pass under or over a road
Ecosystem	An interacting community of independent organisms and the environment they inhabit (English Nature, 1998d)
Ecosystem function	The physical outcome of a species' activity within an ecosystem typically referring to cycling of chemicals or alteration of the physical environment, eg photosynthetic production of oxygen (Jeffries, 1997)
Ecosystem services	The benefits to life, including humanity, accruing from some ecosystems functions (Jeffries, 1997)
Edge	The portion of an ecosystem near its perimeter, where influences of the surroundings prevent development of interior environmental conditions (Forman, 1995)
Edge effect	The distinctive species composition or abundance on an edge
Edge species	Species inhabiting edges or boundaries between biotic communities such as the edge of a woodland (Spellerberg and Sawyer, 1999)

Endemism, endemic	Native, and usually restricted, to a particular geographical region ie it occurs nowhere else but this particular area. Endemism may occur at different levels, subspecies, species, etc.
Enhancement	The genuine enhancement of biodiversity interest eg improved management or new habitats of features, with the result that there is a new benefit to biodiversity ie improvements over and above those required for mitigation/compensation
Environmental factors	All environmental variables that are known to affect organisms; can be divided into abiotic factors, which involve physical and chemical components (eg water, temperature, light, oxygen, nutrients, pH, and toxins) and biotic factors, which involve interactions between organisms (eg competition, predation, parasitism and mutually beneficial relationships such as pollination) (Morris and Therivel, 1995)
Environmental impact statement	Report summarising the findings of an environmental impact assessment. Used interchangeably with environmental statement
Exotic species	A species introduced from one region from another geographical region. Alien species (Spellerberg and Sawyer, 1999)
Fauna	A collective term for all kinds of animals
Flora	A collective term for all kinds of plants
Focusing	The process by which ecological impact assessment is refined, by selecting suitable ecological components for in depth study
Fragility	The inverse of ecosystem stability
Fragmentation	The breaking up of a habitat, ecosystem, or land-use type into smaller parcels. Fragmentation results in the change in the physical environment within the parcels (eg in fluxes of radiation, water and nutrients) and in biogeographic changes (eg in isolation and connectivity) which have important consequences for biota (Gaston and Spicer, 1998)
Gene	A discrete, heritable unit of genetic data, consisting of DNA and carrying the code to regulate a particular characteristic (Jeffries, 1997)
Gene flow	The consequence of cross-fertilisation between members of species across boundaries between populations, or within populations, resulting in the spread of genes across and between populations (Treweek, 1999)
Genetic diversity (variation)	The heritable variation in a population as a result of different variants (the alleles) of any gene (Spellerberg and Sawyer, 1999)
Genotype	The genetic constitution of an organism
Geographical Information Systems (GIS)	Integrated systems of computer hardware and software for entering, storing, retrieving, transforming, measuring, combining, subsetting and displaying spatial data that have been digitised and registered to a common co-ordinate system (Treweek, 1999)
Guiding Principles	The principles which should guide consideration of biodiversity in road EIAs

Guild	A group of species with similar ecological requirements and similar feeding strategies (Spellerberg and Sawyer, 1999)
Habitat	A place in which a particular plant or animal lives. Often used in the wider sense referring to major assemblages of plants and animals found together (English Nature, 1998d)
Habitat capability	The ability of a habitat, under optimal natural conditions to provide life requisites of a species, irrespective of its current habitat conditions (Treweek, 1999)
Habitat Evaluation Procedure	A formal procedure developed by the US Fish and Wildlife Service to assess the consequences of habitat loss for wildlife (Treweek, 1999)
Habitat patch or fragment	A portion of the living space inhabited by populations of species. The habitat patch or fragment is part of a formerly larger area (Spellerberg and Sawyer, 1999)
Habitat potential	A measure of the ability of a given habitat to support a certain species (Treweek, 1999)
Habitat specificity	The degree to which a species is associated with one habitat, compared with its occurrence in all habitats (Treweek, 1999)
Habitat suitability	The ability of a habitat in its current condition to provide life requisites of a species (Treweek, 1999)
Habitat Suitability Index (HSI)	Used in habitat evaluation procedure. Derived by comparing habitat conditions in a study area with optimum conditions for the same evaluation species (Treweek, 1999)
Heterogeneity	The uneven, non-random distribution of objects
Heterozygosity	Genetic variability of individuals and populations of species (Jeffries, 1997)
Home range	The area habitually used by a species to fulfil its requirements for food, shelter and a place to breed. Excursions beyond this area are rare (Treweek, 1999)
Homogeneity	The even distribution of objects
Homozygosity	Genetic uniformity of individuals and populations of species (Jeffries, 1997)
Impact range	The area likely to be affected by a proposed action
Inbreeding	Reproduction within a small population of related individuals, often reducing fitness, (Jeffries, 1997)
Indicator	Any representative component, used to provide surrogate measurements reflecting the likely behaviour of other components (Treweek, 1999)
Indigenous (species)	A species which is native to a particular region (Spellerberg and Sawyer, 1999)
Indirect impact	An impact that is attributable to a defined action or stressor, but that affects an environmental or ecological component via effects on other components. Indirect effects are often, but not necessarily, time-delayed or expressed at some distance from their source (Treweek, 1999)

	Integrity	The coherence of a site's ecological/geological structure and function across its whole area that enables it to sustain the habitat, complex of habitats and/or the levels of populations of the species for which it was designated (English Nature, 1998d)
	Key Objective	To ensure that road schemes: Do not significantly reduce biodiversity at any of its levels; and enhance biodiversity wherever possible
	Keystone species	A species in a community which interacts with other species and upon which many other species depend (Spellerberg and Sawyer, 1999). Also used to describe the effect of a change in one species on some characteristic (eg processes or functions) of its community or ecosystem. Keystone species have an impact that is out of proportion to their proportional abundance (Treweek, 1999). A species on which several other species, or the functioning of an ecosystem, may depend (Morris and Therivel, 1995)
	Landscape	A mosaic where a cluster of local ecosystems is repeated in similar form over a kilometres-wide area (Forman, 1995)
	Landscape element	Each of the relatively homogenous units, or spatial elements recognised at the scale of a landscape mosaic. (This refers to each patch, corridor, and area of matrix in the landscape) (Forman, 1995)
	Matrix	The background ecosystem or land-use type in a mosaic, characterised by extensive cover, high connectivity, and/or major control over dynamics (Forman, 1995)
	Metapopulation	A population perceived to exist as a series of subpopulations linked by migration between them. However, the rate of migration is limited, such that the dynamics of the metapopulation should be seen as the sum of the dynamics of the individual sub populations (Begon et al, 1996)
	Microclimate	The climate of a habitat; a climate affected by the local topography, vegetation, soil, etc. (Spellerberg and Sawyer, 1999)
	Minimum dynamic area	The smallest area required to conserve the totally of patterns, processes and functions of an ecosystem (Jeffries, 1997)
	Minimum viable habitat	The minimum area and quality of habitat required to support a given population (Treweek, 1999)
	Minimum viable population	The smallest isolated population having a 99% chance of remaining in existence for 100 years despite the foreseeable effects of demographic, environmental and genetic stochastically, and natural catastrophes (Treweek, 1999). The smallest isolated population required to ensure a species' survival into the foreseeable future (Jeffries, 1997)
	Mitigation	Measures taken to reduce adverse impacts eg modifications or additions to the design of the development, such as the creation of reed bed silt traps to prevent polluted water running directly into ecologically important watercourses. The preservation of 'wildlife corridors' between habitats which would be separated by a proposed development may reduce the possible effects on some fauna
	Mosaic	A pattern of patches, corridors, and matrix, each composed of small, similar aggregated objects (Forman, 1995)

Lowland

Natural Areas	Biogeographic regions as specified by English Nature
Natural variation	Variation attributable to non-anthropogenic causes
Network	An interconnected system of corridors (Forman, 1995)
Niche	The 'space' or 'ecological role' occupied by a species and the resources used by a species. Conceptually the niche is multidimensional and each resource (food, time of feeding, etc.) and each abiotic factor (salinity, temperature, etc.) can be considered a dimension of the niche (Spellerberg and Sawyer, 1999)
No net loss	The point at which habitat or biodiversity losses equal their gains, both quantitatively and qualitatively (Treweek, 1999)
Outbreeding	Reproduction between individuals not closely related, typically drawn from a large, heterozygous population (Jeffries, 1997)
Patch	A relatively homogeneous non-linear area that differs from its surroundings (Forman, 1995)
Phenotype	The observed characteristics of a species, the result of the genotype interacting with the environment
Population	A collection of individuals (plants or animals), all of the same species and in a defined geographical area (Spellerberg and Sawyer, 1999)
Population density	The numbers in a population per unit area
Population dynamics	The variations in time and space in the size and densities of populations (Treweek, 1999)
Population viability analysis	The structured, systematic and comprehensive examination of the interacting factors that place a population or species at risk (Treweek, 1999)
Precautionary principle	The principle of taking precautionary measures where an activity raises threats or harm to biodiversity even if certain cause and effect relationships are not scientifically established
Project	An individual development scheme
Rarity	A measure of relative abundance
Receptor	Any ecological component affected by a particular action or stressor (Treweek, 1999)
Replaceability	A measure of the extent to which a habitat or ecosystem can be restored or reconstructed (Treweek, 1999)
Riparian	The edge of streams or rivers. Riparian biota is that frequenting or living on the banks of rivers and streams (Spellerberg and Sawyer, 1999)
Resilience	The tendency of a system to return to its former state following a disturbance (Treweek, 1999)

Resource	That which may be consumed by an organism thereby becoming unavailable to other individuals of the same or different species (Treweek, 1999)
Restoration	The re-establishment of a damaged or degraded system or habitat to a close approximation of its pre-degraded condition (Treweek, 1999)
Scale	Spatial proportion, as the ratio on a map to actual length; also the level or degree of spatial resolution perceived or considered. (Fine scale refers to pattern in a small area, where the difference between map size and actual size is relatively low, whereas broad or coarse scale refers to a large area, where the difference is great) (Forman, 1995)
Scoping	Determination of the scope of an EIA
Screening	Determination of whether or not an EIA is necessary
Semi-natural vegetation	Vegetation which has been modified by humans but is still of significant nature conservation interest because it is composed of native plant species, is similar in structure to natural types and supports native animal communities (English Nature, 1998d)
Spatial element	Each of the relatively homogenous units recognised in a mosaic at any scale (Forman, 1995)
Species	A group of organisms of the same kind which reproduce amongst themselves but are usually reproductively isolated from other groups of organisms (Spellerberg and Sawyer, 1999)
Species composition	A qualitative measure of the range of species present (Treweek, 1999)
Species diversity	A measure of species richness and the relative abundance of species (Treweek, 1999)
Species of Conservation Concern	Species on the UK Biodiversity Group's list which fall into one or more of the categories set out on page 7 of this guidance
Species richness	The number of species in an area or a sample (Spellerberg and Sawyer, 1999)
Stability	The ability of an ecosystem to maintain some sort of equilibrium in the presence of perturbations (Treweek, 1999)
Stepping stone	An ecologically suitable patch where an animal temporarily stops while moving along a heterogeneous route
Stochastic processes	Random processes
Succession	The process by which a series of plants colonise a substrate over time, such as a change from open water, through swamp and scrub to woodland (English Nature, 1998d)
Sustainable use	A use which can be continued through time without significantly changing the populations, species and habitats being used (Spellerberg and Sawyer, 1999)
Translocation	The removal and relocation of an individual, a population, a community, or a habitat from one location to another

Trophic level	Position in the food chain
Umbrella species	Those species for which targeted conservation management will also benefit other species using the same habitat (Treweek, 1999)
Vascular plants	All the plants, excluding mosses, liverworts and fungi, etc., and having conducting tissue
Wetland	A biological community in an area of wet ground; areas of marsh, peatland or water whether permanent or temporary, with water which is static or flowing, fresh or brackish. The classification of wetlands is based partly on the types of plant species found there and on the physical characteristics (Spellerberg and Sawyer, 1999)

Reference Boxes (organised by subject area)

International guidance on biodiversity and EIA

- US Council on Environmental Quality (US CEQ) (1993) guidance *Incorporating Biodiversity Considerations Into Environmental Impact Analysis Under the National Environmental Policy Act.* CEQ, Washington, US. Available from the US CEQ website at http://ceq.eh.doe.gov/nepa/nepanet.htm.
- Canadian Environmental Assessment Agency (CEAA) (1996a) guidance *A Guide on Biodiversity and Environmental Assessment.* Minister of Supply and Services, Canada. Available on the CEAA website at http://www.ceaa.gc.ca.
- World Bank Environment Deprtament (1997) Environmental Assessment Sourcebook Update Number 20: *Biodiversity and Environmental Assessment.* The World Bank, Washington DC, US..
- The World Conservation Union (IUCN) is part way through a programme of work for *Addressing Biodiversity Impact Assessment* (IUCN web page – http://www.economics.iucn.org; Bagri *et al*, 1998; Bagri & Vorhies, 1999)
- A framework approach to biodiversity has been discussed by the International Association for Impact Assessment (IAIA) (Sadler, 1996) *International study of the Effectiveness of Environmental Assessment Final Report.* Minister of Supply and Services, Canada.
- At the 18th annual meeting of IAIA (21-22 April 1998) for the first time a workshop was held specifically on biodiversity impact assessment. At the 19th annual meeting of IAIA (15-19 June 1999) a series of workshops on biodiversity impact assessment were held. Abstracts and some of the papers presented at both meetings can be accessed via the IUCN website http://www.economics.iucn.org

Grasshopper

Reference Box 1

Useful environmental/ecological assessment references

- Box, J.D. & Forbes, J.E. (1992) Ecological considerations in the environmental assessment of road proposals, *Highways and Transportation*, April: 16-22.
- DoE (1995) *Preparation of Environmental Statements for Planning Projects that require Environmental Assessment.* HMSO, London.
- Department of Transport (DoT) (1993 onwards – updates and additional guidance issued periodically) *Design manual for Roads and Bridges Volume 11: Environmental Assessment.* HMSO, London. Available on the internet from http://www.official-documents.co.uk/document/ha/dmrb/index.htm.
- English Nature (1994a) *Roads and nature conservation: Guidance on impacts, mitigation and enhancement.* English Nature, Peterborough.
- English Nature (1994b) *Nature conservation in Environmental Assessment.* English Nature, Peterborough.
- English Nature (1995a) *Badgers – guidelines for developers.* English Nature, Peterborough.
- English Nature (1996b) *Great crested newts – guidelines for developers.* English Nature, Peterborough.
- English Nature (1999) Water vole *– guidance for planners and developers.* English Nature, Peterborough.
- Institute of Environmental Assessment (1995) *Guidelines for Baseline Ecological Assessment.* E & FN Spon, London.
- Morris, P. & Therivel, R. (eds.) (1995) *Methods of Environmental Impact Assessment.* UCL Press, London.
- Milko, R (1998a) *Wetlands environmental assessment guideline.* Minister of Public Works and Government Services, Canada. Available on the internet at http://www.cws-scf.ec.gc.ca/eass/intro_e.html.
- Milko, R (1998b) Migratory birds *environmental assessment guideline.* Minister of Public Works and Government Services, Canada. Available on the internet as above.
- Milko, R (1998c) Environmental assessment guideline for forest habitat of migratory birds. Minister of Public Works and Government Services, Canada. Available on the internet as above.
- Petts, J (ed.) (1999a) *Handbook of Environmental Impact Assessment Volume 1 Environmental Impact Assessment: Process, Methods and Potential.* Blackwell Science, Oxford. Especially Chapter 15 by Peter Wathern on ecological impact assessment.
- Petts, J (ed.) (1999b) *Handbook of Environmental Impact Assessment Volume 2 Environmental Impact Assessment in Practice: Impact and Limitations.* Blackwell Science, Oxford. Especially Chapter 16 on environmental impact assessment of road and rail infrastructure.
- RSPB (1995) Wildlife *impact – the treatment of nature conservation in environmental assessment.* The RSPB, Sandy.
- Treweek, J (1996) Ecology and environmental impact assessment, *Journal of Applied Ecology* 33: 191-199.
- Treweek, J (1999) *Ecological Assessment.* Blackwell Science, Oxford.

Reference Box 2

Background biodiversity information

- Heywood, VH (Executive editor) 1997 *Global Biodiversity Assessment.* Cambridge University Press for UNEP, Cambridge.
- Biodiversity Map Library supported by the WCMC
- Wilson, EO (ed.) (1988) *Biodiversity.* National Academy Press, Washington, DC, US.
- Reaka-Kudla, ML, Wilson, DE, & Wilson, EO (eds.) (1997) *Biodiversity II: understanding and protecting our biological resources.* National Academy Press, Washington, DC, US.
- English Nature (1998d) *Natural Areas: nature conservation in context* (CD-ROM). English Nature, Peterborough. This CD ROM has all of the 120 Natural Area profiles each of which sets out biodiversity targets, details of the LEAPs in each Natural Area, and national overviews for some habitat types eg lowland grassland and lowland heath.
- English Nature (1997a) *The character of England: landscape, wildlife and natural features* (CD-ROM). English Nature, Peterborough.
- English Nature *Annual Reports.* Most recently the 7th report covering the year 1 April 1997 – 31 March 1998 (English Nature, 1998a). These reports set out information on SSSIs, NNRs, SPAs, SACs, LNRs, the Species Recovery Programme, and current research projects.
- English Nature's series of regional biodiversity publications *Natural Areas* and *Sustainable Development & Regional Biodiversity Indicators* for the Regions eg English Nature (1999g) *Natural Areas in London and the South East Region.* English Nature, Peterborough and English Nature (1999h) *Sustainable Development & Regional Biodiversity Indicators for London.* English Nature, Peterborough
- The UK BAP- HM Government (1994) *Biodiversity: the UK Action Plan* (Command 2428). HMSO, London.
- Wynne, G, Avery, M, Campbell, L, Gubbay, S, Hawkswell, S, Juniper, T, King, M, Newbery, P, Smart, J, Steel, C, Stones, T, Stubbs, A, Taylor, J, Tydeman, C, & Wynde, R (1995) *Biodiversity Challenge: an agenda for conservation in the UK (second edition).* The RSPB, Sandy.
- National HAPs and SAPs (HM Government (1995b) *Biodiversity: the UK Steering Group Report Volume 2: Action Plans.* HMSO, London; English Nature (1998b) *UK Biodiversity Group: Tranche 2 Action Plans Volume I – vertebrates and vascular plants.* English Nature, Peterborough; English Nature (1998c) *UK Biodiversity Group: Tranche 2 Action Plans Volume II – terrestrial and freshwater habitats.* English Nature, Peterborough; English Nature (1999a) *UK Biodiversity Group: Tranche 2 Action Plans Volume III– Plants and Fungi.* English Nature, Peterborough; English Nature (1999b) *UK Biodiversity Group: Tranche 2 Action Plans Volume IV– Invertebrates.* English Nature, Peterborough; English Nature (1999c) *UK Biodiversity Group: Tranche 2 Action Plans Volume V – marine species and habitats.* English Nature, Peterborough; English Nature (1999d) *UK Biodiversity Group: Tranche 2 Action Plans Volume VI – terrestrial and freshwater species and habitats.* English Nature, Peterborough). These set out the current status and current factors affecting the habitat/species, current action, action plan objectives and proposed targets and proposed actions. Details of the national HAPs and SAPs likely to be threatened by road developments are given in Appendices 2 and 3.
- National HSs (HM Government, 1995; English Nature, 1998c and d). These set out the current status and current factors affecting the habitat, current action, and the conservation direction for the habitat. As the broad habitat classification has been revised some of the HSs are out of date. A detailed interpretation manual is being prepared by JNCC and should be available shortly.
- English Nature (1999i) *Biodiversity: Making the links.* English Nature, Peterborough which identifies associations between species and habitats for which BAPs have been prepared.

Reference Box 3

Background biodiversity information (cont)

- Regional Biodiversity Audits eg RSPB & SW Regional Planning Conference (edited by Cordery, L) (1996) *The Biodiversity of the South west: an audit of the South-West biological resource*. The RSPB, Sandy; RSPB (1999) *A Biodiversity Audit of North West England (volumes 1 and 2)*. The RSPB, Sandy; Hampshire & Isle of Wight Wildlife Trust (1999) *The Biodiversity of South East England: An Audit and Assessment*.; Selman, R, Dodd, F and Baynes, K (1999) *A Biodiversity Audit of Yorkshire and the Humber*, Yorkshire and Humber Biodiversity Forum.
- LBAPs. Consult the UK Biodiversity Secretariat database - DETR (1999a) *List of LBAPs and contacts. DETR, Bristol*. Available on The UK Biodiversity Secretariat's website at http://www.jncc.gov.uk/ukbg - to find the appropriate LBAP(s).
- Local Records Centres are often importance holders of local biodiversity data eg County surveys, species surveys, details of local specialist groups.
- *Biodiversity News* the quarterly newsletter of the UK Biodiversity Secretariat. Available from the Secretariat and on the Secretariat's website.
- Company BAPs eg Wessex Water (1999) *Wessex Water Biodiversity Action Plan*. Wessex Water; North West Water Ltd (1999) *North West Water – From Rio to Rivington*. North West Water.; Northumbrian Water (1999) *Northumbrian Water Biodiversity Strategy*. Northumbrian Water, Durham.

Reference Box 3 (cont)

Biodiversity websites

- There are numerous websites with biodiversity information. Key sites include:
- *The UK National Biodiversity Network (NBN)* website at http://www.nbn.org.uk/. The NBN aims to provide a publicly accessible database on environmental data including British biodiversity which links the use of wildlife data to its collection. The NBN is currently in its early stages.
- *The UK Biodiversity Group/ UK Biodiversity Secretariat* website at http://www.jncc.gov.uk/ukbg. This site includes the LBAP database, the national HAPs and SAPs, and the biodiversity Secretariat's newsletter.
- *The Scottish Biodiversity Group (SBG)* website at http://www.scotland.gov.uk/biodiversity.
- *The Welsh Biodiversity Group* website at http://www.ccw.gov.uk/biodiv.
- *Clearing House Mechanism of the CBD (CHM)* – http://biodiv.org.
- *UK Clearing House Mechanism* – http://www.chm.org.uk
- *Biodiversity Conservation Information System (BCIS)* website at http://biodiversity.org provides a guide to available biodiversity information.
- *Bionet* website at http://www.igc.apc.org/bionet/. Bionet (the Biodiversity Action Network) is a non-governmental-organisation network that aims to strengthen biodiversity law and policy and inform the environmental community and others about biodiversity issues.
- *The International Institute for Sustainable development (IISD)* webpage at http://www.iisd.ca/linkages/biodiv/biodivsites.html has good links to other useful biodiversity websites.
- *County Wildlife Trust websites* at http://www.wildlifetrust.org.uk/ eg Cornwall Wildlife Trust website at http://www.wildlifetrust.org.uk/cornwall which includes a complete biodiversity audit and is good for localised information.
- *Biodiversity Partnership/Initiative websites* eg the website of the Nottingham Biodiversity Action Group at http://www.nottsbag.org.uk.
- *The Pan-European Biological and Landscape Diversity Strategy* website at http://www.strategyguide.org/. This site provides an information, communication and monitoring programme in support of the Pan-European strategy.
- *The EIONET* website at http://www.eionet.eu.int/ec-chm/index.html. EIONET (European Environment Agency's Information and Observation Network) is developing an EC clearing house mechanism.

Reference Box 4

Key scientific references

- Bibby, CJ, Burgess, ND & Hill, DA (1992) *Bird Census techniques, BTO and RSPB, Academic Press*, London.
- Byron, HJ, Treweek, JR, Sheate, WR & Thompson, S. (2000) Road developments in the UK: an analysis of ecological assessment in environmental impact statements produced between 1993 and 1997. *Journal of Environmental Planning and Management,* 43(1), 71-97.
- Canters, K, Piepers, A, & Hendriks-Heersma, D (eds.) (1995) *Proceedings of the international conference on habitat fragmentation, infrastructure and the role of ecological engineering,* 17-21 Sept 1995, Maastricht and the Hague.
- English Nature (1994a) *Roads and nature conservation: Guidance on impacts, mitigation and enhancement,* English Nature, Peterborough
- English Nature (1996a) *The significance of secondary effects from roads and road transport on nature conservation - English Nature Research Reports No 178,* English Nature, Peterborough.
- English Nature (1999e) *List of English Nature Research Reports,* English Nature, Peterborough. Particular reports may be relevant to individual EIAs eg Report No 298 *Invertebrates and their habitats in Natural Areas. Vol. 1: Midland & Northern Areas. Vol. 2: Southern Areas* and Report No 275 *The area of key habitats in the East Anglian Plain.*
- Evink, GL, Garrett, P, Zeigler, D & Berry, J (eds.) (1996) *Trends in addressing transportation related wildlife mortality Proceedings of the Transportation Related Wildlife Mortality Seminar.* State of Florida, Department of Transportation, Environmental Management Office, June 1996.
- Evink, GL, Garrett, P, Zeigler, D & Berry, J (eds.) (1998) *Proceedings of the International Conference On Wildlife Ecology and Transportation,* February 10-12, 1998, Ft. Myers, Florida. State of Florida, Department of Transportation, Environmental Management Office.
- Evink, GL, Garrett, P, Zeigler, D & Berry, J (eds.) (1999) *Proceedings of the International Conference On Wildlife Ecology and Transportation,* September 13-16, 1999, Missoula, Montana. State of Florida, Department of Transportation, Environmental Management Office.
- Forman, R.T.T. (1995) *Land Mosaics: The ecology of landscapes and regions,* Cambridge University Press, Cambridge.
- Gibbons, D, Avery, M, Baillie, S, Gregory, R, Kirby, J, Porter, R, Tucker, G & Williams, G (1996) Bird species of conservation concern in the United Kingdom, Channel Islands and the Isle of Man: revising the red data list, *RSPB Conservation Review 10: 7-18,* RSPB, Sandy.
- IENE (Infra Eco Network Europe) Proceedings see – http://iene.vv.se. IENE has started a EU COST (Co-operation in the field of Scientific and Technical Research) action 341 *Habitat fragmentation due to transport infrastructure,* see webpage - http://iene.vv.se/coordcost.htm
- Linnean Society (Ed) (2000) *Proceedings of a Linnean Society/RSPB/WWF-UK joint symposium Wildlife and Roads: The ecological impact,* London, 11-12 March 1998, Imperial College Press (in press).
- RSPB (1998) *Land For Life: Technical Support Document,* RSPB, Sandy. (RSPB's analysis of impacts on SSSIs) This report includes information on the extent and changes in area of key habitats and sites and species case studies for a range of biodiversity, indicating changes in status and the relevant conservation issues.
- Reijnen, MJSM, Veenbas, G & Foppen, RPB (1995) *Predicting the effects of motorway traffic on breeding bird populations,* Road and Hydraulic Engineering Division and DLO-Institute for Forestry and Nature Research, Delft, The Netherlands.
- Spellerberg, I.F. (1998) Ecological effects of roads and traffic: a literature review, *Global Ecology and Biogeography Letters* 7: 317-333.
- Spellerberg, I.F. & Morrison, T. (1998) The ecological effects of new roads - a literature review, *Science for Conservation, 84,* Department of Conservation, Wellington, New Zealand.
- Sutherland, WJ (1996) *Ecological census techniques,* Cambridge University Press, Cambridge.

Reference Box 5

Key scientific references (cont)

- Treweek, JR, Thompson, S, Veitch, N & Japp, C (1993) Ecological assessment of proposed road developments: a review of environmental statements, *Journal of Environmental Planning & Management*, 36: 295-307.
- Thompson, S, Treweek, JR, & Thurling, DJ (1997) The ecological component of environmental impact assessment: a critical review of British Environmental statements, *Journal of Environmental Planning & Management*, 40(2): 157-171.
- Tucker, GM & Heath, MF (eds.)(1994) *Birds in Europe: their conservation status.* BirdlLife Conservation Series No. 3, Cambridge.
- Wenatchee Forestry Sciences Lab (1999) *Wildlife linkage assessment project comprehensive bibliography of the published literature on roadway and wildlife interactions,* can be downloaded from http://www.fs.fed.us/pnw/wenlab/research/projects/wildlife/index.html.
- Wildlife Trusts *Head on Collision Report Series* (Regional analyses of impacts of roads on wildlife) Eg Scottish Wildlife Trust (1996) *Head on Collission Scotland*, Scotlish Wildlife Trust, Edinburgh, and The Wildlife Trusts - South East England (1994) *Head on Collision 1994*, The Wildlife Trusts - South East England.
- South East England (1994) *Head on Collision 1994,* The Wildlife Trusts - South East England.

Reference Box 5 (cont)

Biodiversity measurement references

- Gaston, KJ (ed.) (1996) *Biodiversity: A Biology of Numbers and Difference*, Blackwell Science, Oxford.
- HMSO (1996a) *Biodiversity Assessment: A Guide to good practice Vol. 1*, HMSO, London.
- HMSO (1996b) *Biodiversity Assessment: A Guide to good practice Vol. 2 (Field manual 1 data and specimen collection of plants, fungi and micro-organisms)*, HMSO, London.
- HMSO (1996c) *Biodiversity Assessment: A Guide to good practice Vol. 3 (Field manual 2 data and specimen collection of animals)*, HMSO, London.
- Hawksworth, D.L. (Ed) (1996) *Biodiversity: measurement and estimation*, Chapman Hall, London.
- Noss, R.F. (1990) Indicators for monitoring biodiversity: a hierarchical approach,

Reference Box 6

Cliffs

Useful mitigation references

- Aanen *et al* (1991) *Nature Engineering and Civil Engineering Works*. Ministry of Transport and Public Works, Directorate-General for Public Works and Water Management, Road and Hydraulic Engineering Division, The Netherlands.
- Canters *et al* (eds.) (1995) *Proceedings of the international conference on habitat fragmentation, infrastructure and the role of ecological engineering*, 17-21 Sept 1995, Maastricht and the Hague.
  Maastricht conference.
- Cuperus, R Canters, KJ, Udo de Haes, HA, & Friedman, DS (1999) Guidelines for ecological compensation associated with highways. *Biological Conservation*, 90: 41-51.
- English Nature (1994a) *Roads and nature conservation: Guidance on impacts, mitigation and enhancement.* English Nature, Peterborough.
- English Nature (1994b) *Nature Conservation in Environmental Assessment.* English Nature, Peterborough.
- English Nature (1996b) *Great crested newts – guidelines for developers.* English Nature, Peterborough.
- English Nature (1995a) *Badgers – guidelines for developers.* English Nature, Peterborough.
- English Nature (1999f) Water vole *– guidance for planners and developers.* English Nature, Peterborough.
- Linnean Society (Ed) (2000) *Proceedings of a Linnean Society/RSPB/WWF-UK joint symposium Wildlife and Roads: The ecological impact*, London, 11-12 March 1998, Imperial College Press (in press).
- Reijnen *et al* (1995) *Predicting the effects of motorways on breeding bird populations.* Ministry of Transport and Public Works, Directorate-General for Public Works and Water Management, Road and Hydraulic Engineering Division and DLO-Institute for Forestry and Nature research, The Netherlands.
- Road and Hydraulic Engineering Division (1993) *Prediction methods - Environmental Impacts - Infrastructural Projects*. Ministry of Transport and Public Works, Directorate-General for Public Works and Water Management, Road and Hydraulic Engineering Division, The Netherlands.
- Road and Hydraulic Engineering Division (1995a) *Nature across motorways*. Ministry of Transport and Public Works, Directorate-General for Public Works and Water Management, Road and Hydraulic Engineering Division, The Netherlands.
- Road and Hydraulic Engineering Division (1995b) *Wildlife Crossings for Roads and Motorways*. Ministry of Transport and Public Works, Directorate-General for Public Works and Water Management, Road and Hydraulic Engineering Division, The Netherlands.

Reference Box 7

Useful mitigation references (cont)

- Road and Hydraulic Engineering Division (1995c) *Dispersal of animals and infrastructure – A model study: Summary.* Ministry of Transport and Public Works, Directorate-General for Public Works and Water Management, Road and Hydraulic Engineering Division, The Netherlands.
- Road and Hydraulic Engineering Division (1995d) *Mammal use of fauna passages on national road A1 at Oldenzaal.* Ministry of Transport and Public Works, Directorate-General for Public Works and Water Management, Road and Hydraulic Engineering Division, The Netherlands.
- Treweek, J & Thompson, S (1997) A review of ecological mitigation measures in UK environmental statements with respect to sustainable development, *International Journal of Sustainable Development and World Ecology,* 4: 40-50.
- *infrastructure – A model study: Summary.* Ministry of Transport and Public Works, Directorate-General for Public Works and Water Management, Road and Hydraulic Engineering Division, The Netherlands.
- Road and Hydraulic Engineering Division (1995d) *Mammal use of fauna passages on national road A1 at Oldenzaal.* Ministry of Transport and Public Works, Directorate-General for Public Works and Water Management, Road and Hydraulic Engineering Division, The Netherlands.
- Treweek, J & Thompson, S (1997) A review of ecological mitigation measures in UK environmental statements with respect to sustainable development, *International Journal of Sustainable Development and World Ecology,* 4: 40-50.

Reference Box 7 (cont)

Habitat creation and translocation references

- Bullock, JM (1998) Community translocation in Britain: setting objectives and measuring consequences. *Biological Conservation,* 84, 199-214.
- English Nature (1997b) *Research report No. 260 – Habitat restoration Project: Factsheets and bibliographies.* English Nature, Peterborough.
- English Nature (1995b) *Habitat creation – a critical guide.* English Nature, Peterborough.
- English Nature (1998e) *Research report No. 269 – The restoration of replanted ancient woodland.* English Nature, Peterborough.
- Gault, C (1997) *A Moving Story – Species and community translocation in the UK: a review of policy, principle, planning and practice.* WWF-UK, Godalming.
- Gilbert, OL & Anderson, P (1998) *Habitat creation and repair.* Oxford University Press, Oxford.
- Institute of Terrestrial Ecology (ITE) (1997) *Ecology and Twyford Down.* ITE, Furzebrook Research Station, Wareham.

Reference Box 8

General references

Association of Local Government Ecologists (ALGE) and the South West Biodiversity Partnership (2000) *A Biodiversity Guide for the Planning and Development Sectors.* RSPB, Exeter.

Bagri, A, McNeely, J & Vorhies, F (1998) *Biodiversity and Impact Assessment.* IUCN, Gland Switzerland. Available on http://economics.iucn.org/themes-a.htm.

Barnes, JL & Davey, LH (1999) *A Practical Approach to Integrated Cumulative Environmental Effects Assessment to Meet the Requirements of the Canadian Environmental Assessment Act,* prepared for a workshop on Cumulative Effects Assessment, 19th Annual Meeting of IAIA, Glasgow, Scotland, 16-17 June 1999.

Beanlands, GE & Duinker, PN (1983) *An ecological framework for environmental impact assessment in Canada.* Institute for Resource and Environmental Studies, Dalhousie University, Halifax; in co-operation with the Federal Environmental Impact Assessment Review Office, Canada.

Beanlands, GE & Duinker, PN (1984) An ecological framework for environmental impact assessment. *Journal of Environmental Management* 18: 267-277.

Begon, M, Harper, JL & Townsend, CR (1996) *Ecology: Individuals, Populations and Communities,* 3rd edition. Blackwell Science, Oxford.

Blue Circle (1997) *Medway Works Environmental Statement,* Volumes 1-3, December 1997. Blue Circle.

Buckley, RC (1991) How accurate are impact predictions? *Ambio* 20: 161-162.

Byron, HJ & Sheate, WR (2000) Treatment of biodiversity issues in road environmental impact assessments, in *Proceedings of a Linnean Society/RSPB/WWF-UK joint symposium Wildlife and Roads: The ecological impact,* London, 11-12 March 1998, Imperial College Press (in press).

Byron, HJ, Treweek, JR, Sheate, WR & Thompson, S. (2000) Road developments in the UK: an analysis of ecological assessment in environmental impact statements produced between 1993 and 1997. *Journal of Environmental Planning and Management,* 43(1), 71-97.

Canter, L.W (1996) *Environmental Impact Assessment,* second edition. McGraw-Hill International, New York.

Commission of the European Communities (CEC) (1985) Council Directive 85/337/EEC on the assessment of the effects of certain public and private projects on the environment. *Official Journal of the European Communities,* L175, 5.7.1985, pp40-48.

CEC (1996) *Evaluation of the performance of the EIA process,* Wood, C, Barker, A, Jones, C & Hughes, J (eds.). Volumes 1 and 2, October 1996.

CEC (1997) Council Directive 97/11/EC. *Official Journal of the European Communities,* L73, 14.3.97, p5.

DeLong, Jr, DC (1996) Defining biodiversity, *Wildlife Society Bulletin* 1996, 24(4): 738-749.

Department of the Environment(DoE)/Welsh Office (1989) *Environmental Assessment: A Guide to the Procedures.* HMSO, London.

DoE (1994) *Planning Policy Guidance Note (PPG) 9: Nature Conservation.* HMSO, London.

DoE (1995) *Preparation of Environmental Statements for Planning Projects that require Environmental Statements.* HMSO, London

DoE-Northern Ireland (1989) *Development Control Advice Note 10 – Environmental Assessment.* HMSO, London.

DoE-Northern Ireland (1996) *Planning Policy Statement (PPS) 3 Development Control: Roads Considerations.* HMSO, London.

DoE-Northern Ireland (1997) *Planning Policy Statement (PPS) 2 Planning and Nature Conservation.* HMSO, London.

Department of Environment, Transport and the Regions (DETR) (1997) *Mitigation measures in Environmental Statements.* DETR, London.

DETR (1998a) *Sustainable Development: Opportunities for Change – Making Biodiversity Happen.* DETR, London.

DETR (1998b) *A New Deal for Trunk Roads in England.* DETR, London.

DETR (1998c) *A New Deal for Trunk Roads in England: Guidance on the New Approach to Appraisal.* DETR, London.

DETR (1998d) *A New Deal for Trunk Roads in England: Understanding the New Approach to Appraisal.* DETR, London.

DETR (1999a) *List of LBAPS and Contacts.* DETR Bristol.

DETR (1999b) *Circular 02/99 on the Town and Country Planning (Environmental Impact Regulations) (England and Wales) Regulations 1999 No. 293.* DETR, London.

DETR (2000) *Guidance on the Methodology for Multi-Modal Studies (GOMMMS).* DETR, London.

Department of Transport (DoT) (1993) *Design manual for roads and bridges volume 11: Environmental Assessment* (together with subsequent additions) HMSO, London.

DoT (1994) M25 Motorway Link Roads between junctions 12 and 15 Environmental Statement. DoT, London.

DiSilvestro, RL (1993) *Reclaiming the last wild places: a new agenda for biodiversity.* John Wiley, New York & Chichester.

Donnelly, A , Dalal-Clayton, D & Hughes, R (1998) *A directory of impact assessment guidelines*, 2nd edition. International Institute for Environment and Development, London.

Duinker, PN (1987) Forecasting environmental impacts: better quantitative and wrong than qualitative and untestable! In Sadler (ed.) (1987) *Audit and evaluation in environmental assessment and management: Canadian and international experience,* Volume II: 399-407. Beauregard Press Ltd, Canada.

English Nature (1993) *Position Statement on Sustainable development.* English Nature, Peterborough. Reviewed and reprinted in April 1999.

English Nature (1994a) Roads and nature conservation: Guidance on impacts, mitigation and enhancement, English Nature, Peterborough

English Nature (1994b) *Nature conservation in Environmental Assessment.* English Nature, Peterborough.

English Nature (1997a) *The character of England: landscape, wildlife and natural features* (CD-ROM). English Nature, Peterborough.

English Nature (1998a) *Annual Report for the year 1 April 1997 – 31 March 1998.* English Nature, Peterborough.

English Nature (1998d) *Natural Areas: nature conservation in context* (CD-ROM). English Nature, Peterborough.

Environment Agency (1998) *National Environmental Assessment Handbook, Environment Agency Internal Works and Activities.*

Environment Agency (Anglian Region) (1999) *Environmental Action Plans: Good Practice Guidelines.* Draft produced for consultation.

European Commission (2000) *Managing Natura 2000 sites: The provisions of Article 6 of the 'Habitats' Directive 92/43/EEC.* European Commission Brussels

Forman, RTT & Deblinger, RD (1998) *The Ecological Road-Zone Effect for Transportation Planning and Massachusetts Highway Example* in Evink, GL, Garrett, P, Zeigler, D & Berry, J (eds.) (1998) *Proceedings of the International Conference On Wildlife Ecology and Transportation,* February 10-12, 1998, Ft. Myers, Florida. State of Florida, Department of Transportation, Environmental Management Office.

Gaston, KJ & Spicer, JI (1998) *Biodiversity: An introduction.* Blackwell Science, Oxford.

Geraghty, P (1999) *A comparative study of guidance documents for EIA and their potential for supporting practice,* prepared for a workshop, 19th Annual Meeting of IAIA, Glasgow, Scotland, 15-19 June 1999.

Hatton, C (2000) Requirements of UK and EU Legislation, in *Proceedings of a Linnean Society/RSPB/WWF-UK joint symposium Wildlife and Roads: The ecological impact,* London, 11-12 March 1998, Imperial College Press (in press).

HM Government (1994) *Biodiversity: The UK Action Plan.* HMSO, London.

HM Government (1995a) *Biodiversity: the UK Steering Group Report Volume 1: Meeting the Rio challenge.* HMSO, London.

HM Government (1995b) *Biodiversity: The UK Steering Group Report Volume 2: Action Plans.* HMSO, London.

Highways Agency (1994) *A1 Motorway North of Leeming to Scotch Corner Environmental Statement.* Highways Agency, Yorkshire and Humberside Construction Programme Division, April 1994.

Highways Agency (1997) A249 *Iwade Bypass to Queenborough Improvement Environmental Statement.* Highways Agency, Southern Operations Division, January 1997.

Hirsch, A (1993) Improving Consideration of Biodiversity in NEPA Assessments. *The Environmental Professional,* Volume 15(1) 103-115.

Institute of Environmental Assessment (1995) *Guidelines for Baseline Ecological Assessment.* E & FN Spon, London.

International Association of Impact Assessors (1998) *Statement on Environmental Assessment and Biodiversity.* IAIA, April 1998. Available on the IUCN website at http://economics.iucn.org.

Jacques Whitford Environment Limited (1993) *Environmental Evaluation of Strait Crossing Inc.'s proposed Northumberland Strait Crossing Project.* Prepared for Strait Crossing Inc., Calgary, Canada, 22 April 1993.

Jeffries, MJ (1997) *Biodiversity and conservation.* Routledge, London & New York.

Joint Nature Conservation Committee (JNCC) (1993) *Handbook for Phase 1 habitat survey: A technique for environmental audit.* JNCC, Peterborough.

JNCC (1998) *Common standards for monitoring designated sites.* JNCC, Peterborough.

Kent Biodiversity Action Plan Steering Group (1997) *The Kent Biodiversity Action Plan: A framework for the future of Kent's wildlife.* Kent County Council, Maidstone.

Le Maitre, D, Euston Brown DIW & Gelderblom CM (1997) *Are the potential Impacts on biodiversity Adequately Addressed in Southern Africa Environmental Impact Assessments?* CSIR, Programme and Papers for the IAIAsa 1997 Conference on 'Integrated Environmental Management in Southern Africa: the state of the art and lessons learnt' (Compiled by G. Kruger), pp 173-182, KwaMaritane, South Africa. Available on the IUCN website at http://economics.iucn.org.

Manchester University EIA Centre (1999) Workshop on the Role of monitoring and post-auditing in the EIA Process held on 13 April 1999 at EIA Centre, Manchester. Papers should shortly be available on the EIA Centre website at http://www.art.man.ac.uk/eia/eiac.htm.

Morris, P. & Therivel, R. (eds.) (1995) *Methods of Environmental Impact Assessment.* UCL Press, London.

Nature Conservancy Council (NCC) (1989) *Guidelines for selection of biological SSSIs.* NCC, Peterborough.

Noss, RF (1990) Indications for monitoring biodiversity: a hierarchical approach, *Conservation Biology,* 4, 355-364.

Prendergast JR & Eversham BC (1997) Species covariance in higher taxa: empirical tests of the biodiversity indicator concept, *Ecography* Volume 20 No. 1 pp210-216.

Rackham, O (1986) *The History of the Countryside.* Dent, London.

Ratcliffe, DA (ed.) (1977) *A Nature Conservation Review.* Cambridge University Press, Cambridge.

Reid, WV, McNeely, JA, Tunstall, DE, Bryant, DA & Winograd, M (1993) *Biodiversity Indicators for Policy-makers.* World Resources Institute, October 1993.

RSPB (1995) Wildlife impact – the treatment of nature conservation in environmental assessment. The RSPB, Sandy.

RSPB (1996) *Sustainable development: the importance of biodiversity.* The RSPB, Sandy.

Royal Town Planning Institute (RTPI) (1999) *Planning for Biodiversity: Good practice guide.* RTPI, London.

Sadler (1996) *International study of the Effectiveness of Environmental Assessment Final Report.* Minister of Supply and Services, Canada.

Scottish Executive (1999a) *Trunk Road Biodiversity Action Plan: Review for Discussion.* Scottish Executive, Edinburgh.

Scottish Executive (1999b) *Circular 15/99 on the Environmental Impact Assessment (Scotland) Regulations 1999 (Scottish SI 1999 No. 1).* Scottish Executive Development Department, Edinburgh.

Scottish Executive (1999c) *Planning Advice Note (PAN) 58 - Environmental Impact Assessment.* Scottish Executive Development Department, Edinburgh.

Scottish Office Environment Department (SOED) (1995) *Circular 6/1995, Nature Conservation: Implementation in Scotland of EC Directives on the Conservation of Natural Habitats and of Wild Flora and Fauna, and the Conservation of Wild Birds: The Conservation (Natural Habitats Etc.) Regulations 1994.* Scottish Office Agriculture, Environment and Fisheries Department.

SOED (1999) *National Planning Policy Guidance (NPPG 14) Natural Heritage.* Scottish Office Agriculture, Environment and Fisheries Department.

Spellerberg, IF (1991) *Monitoring ecological change.* Cambridge University Press, Cambridge.

Spellerberg, I.F. & Morrison, T. (1998) The ecological effects of new roads - a literature review, *Science for Conservation, 84,* Department of Conservation, Wellington, New Zealand.

Spellerberg, IF & Sawyer, JWD (1999) *An introduction to applied biogeography.* Cambridge University Press, Cambridge.

Southerland, MT (1995) Conserving Biodiversity in Highway Development Projects. *The Environmental Professional,* Volume 17 pp. 226-242.

Surrey County Council (1995) *A322 Improvement – Bisley Common to Brookwood Crossroads Environmental Statement .* Surrey County Council, Surrey.

Takacs, D (1997) *The idea of biodiversity: philosophies of paradise.* The John Hopkins University Press, Baltimore and London.

Thames Water (1998) *Planning for Future Water Resources. Best Practicable Environmental Option Study: Methodology for Consultation. No. 1 Report.* Prepared by Land Use Consultants and Consultants in Environmental Services.

Treweek, JR, Thompson, S, Veitch, N & Japp, C (1993) The ecological component of environmental impact assessment: a critical review of British Environmental statements. *Journal of Environmental Planning & Management,* 40(2): 157-171.

Treweek, J & Veitch, N (1996) The potential application of GIS and remotely sensed data to the ecological assessment of proposed new road schemes. *Global Ecology and Biogeographical Letters* 5: 249-257.

Treweek, J (1999) *Ecological Assessment.* Blackwell Science, Oxford.

UNCED (United Nations Commission on Environment and Development) (1992) *Convention on Biological Diversity.* UNCED, Rio.

UNEP (United Nations Environment Programme) (1997a) *Recommendations for a core set of indicators of biological diversity.* UNEP/CBD/SBSTTA/3/9 dated 10 July 1997. Available on the CBD website at http://www.biodiv.org/.

UNEP (1997b) *Recommendations for a core set of indicators of biological diversity: Background paper prepared by the liaison group on indicators of biodiversity.* UNEP/CBD/SBSTTA/3/Inf.13 dated 22 July 1997. Available on the CBD website at http://www.biodiv.org/.

UNEP (1997c) *Indicators of forest biodiversity*. UNEP/CBD/SBSTTA/3/Inf.23 dated 15 August 1997. Available on the CBD website at http://www.biodiv.org/.

UNEP (1998a) *Impact Assessment and Minimising adverse Impacts: Implementation of Article 14*. UNEP/CBD/COP/4/2. Available on the CBD website at http://www.biodiv.org/.

UNEP (1998b) *Measures for Implementing the Convention on Biological Diversity The Fourth Conference of the Parties (COP4) Decision IV/10*. Available on the CBD website at http://www.biodiv.org/.

US Army Corp of Engineers (1990) *A Habitat Evaluation System for Water Resources Planning*. US Army Corp of Engineers, Lower Mississippi Valley Division, Vicksburg, Mississippi.

US Council on Environmental Quality (US CEQ) (1993) *Incorporating Biodiversity Considerations Into Environmental Impact Analysis Under the National Environmental Policy Act.* CEQ, Washington, US.

US Fish and Wildlife Service (1980) *Habitat Evaluation Procedures (HEP)*. ESM 102, US Fish and Wildlife Service, Washington, DC.

Usher, MB (ed.) (1986) *Wildlife Conservation Evaluation*. Chapman and Hall, London.

Warwickshire County Council (1994) *A452 Leamington-Kenilworth Road Improvement Scheme Environmental Statement*. Warwickshire County Council Department of Planning, Transport & Economic Strategy, November 1994.

Warwickshire County Council (1996) *A429 Barford Bypass Environmental Statement*. Warwickshire County Council, June 1996.

Welsh Office (1996) *Planning guidance (Wales) Technical Advice Note 5: Nature Conservation and Planning*. Welsh Office, Cardiff.

Welsh Office Highways Directorate (1997) *A465 Abergavenny to Hirwaun Dualling Environmental Statement*, 31 October 1997. Prepared by Babtie Group Ltd. Welsh Office Highways Directorate, Cardiff.

World Bank Environment Department (1997) Environmental Assessment Sourcebook Update Number 20: *Biodiversity and Environmental Assessment*. The World Bank, Washington DC, US.

The World Bank Environment Department (1999) The evolution of environmental assessment in the World Bank: from 'Approval' to Results. *Environmental Department Papers, Environmental management Series, Paper No. 67. The* World Bank, Washington DC, US.

Wynne, G (ed.) (1993) *Biodiversity Challenge: an agenda for conservation in the UK (first edition)*. The RSPB, Sandy.

Wynne, G, Avery, M, Campbell, L, Gubbay, S, Hawkswell, S, Juniper, T, King, M, Newbery, P, Smart, J, Steel, C, Stones, T, Stubbs, A, Taylor, J, Tydeman, C, & Wynde, R (1995) *Biodiversity Challenge: an agenda for conservation in the UK (second edition)*. The RSPB, Sandy.

Appendix 1 – Key provisions of existing wildlife policy and legislation

Policy/legislation	Objective	Key provisions	Current position
The Ramsar Convention 1971 (The Convention on Wetlands of International Importance Especially as Waterfowl Habitat)	To provide a framework for international co-operation for the conservation and wise use of wetlands and their resources	• Designation of wetlands of international importance as Ramsar sites • Inclusion of wetland conservation in national land use planning	• All UK Ramsar sites are also SSSIs and many are also SPAs • In March 1998 the UK had listed 120 Ramsar sites covering a total of 491 646 hectares (Hatton, 2000)
The Birds Directive 1979 (EC Council Directive on the Conservation of Wild Birds 79/409/EEC)	'…the conservation of all species of naturally occurring birds in the wild state in the European territory of the Member States' (Article 1(1))	• A general level of protection for all wild birds in the territory of the EC • Designation of Special Protection Areas (SPAs) to conserve the habitat of certain particularly rare species and of migratory species listed in Annex I	• All UK SPAs are also SSSIs • In March 1998 the UK had classified 169 SPAs covering over 708 890 hectares (Hatton, 2000). There are 175 species listed in Annex I. Those in the UK include the Whooper swan, corncrake and stone curlew
The Habitats Directive 1992 (EC Council Directive on the Conservation of Natural Habitats and of Wild Fauna and Flora 92/43/EEC)	'….to contribute towards ensuring biodiversity through the conservation of natural habitats and of wild flora and fauna' in the EC (Article 2(1))	• Establishment of Special Areas of Conservation (SAC) to maintain at (or restore to) 'favourable conservation status' (FCS) the habitats and species of Community importance listed in Annexes I and II. Certain habitats and species are identified as priority • FCS is defined as 'when the species population and range is stable (or increasing) and there is a sufficiently large area of habitat available to maintain its population on a long-term basis' (Article 1(j)) • Protection of species outside SACs	• Implemented in the UK by The Conservation (Natural Habitats, &c.) Regulations 1994 • At March 1998 the UK Government had sent a list of 262 potential SACs covering 1 526 817 hectares to the European Commission • 75 of the 168 Annex I habitats occur in the UK including raised bogs, and old oak woodlands with holly and hard fern • Priority habitats in the UK include active raised bogs, and lagoons • 40 of the 193 animal and 432 plant species listed in Annex II occur in the UK including the otter, stag beetle, fen orchid and shore dock • The only priority species in the UK is the Western rustwort

Black Grouse

Policy/ legislation	Objective	Key provisions	Current position
Natura 2000	To establish a network of protected areas as a coherent European ecological network (Article 3(1) Habitats Directive). Together SPAs and SACs will make up Natura 2000	• Management plans where appropriate (Article 6(4) Habitats Directive) • An environmental assessment of all non-management projects which may have a significant effect on Natura 2000 sites to evaluate whether it will affect FCS of the relevant habitat/species (Article 6(3) Habitats Directive). If there will be a negative ecological impact the project may only proceed if certain conditions are satisfied • Where a project does go ahead compensatory measures must be taken to ensure that the overall coherence of Natura 2000 is protected (Article 6(4)	• SPA boundaries must reflect only ecological factors: ECJ Case 44/95 R v SoS for the Environment ex parte RSPB, judgement delivered 11/7/96 and R v SoS for the Environment ex parte RSPB Judgement Order of the House of Lords delivered 13/3/97 • Will an EIA be satisfactory as an Article 6(3) assessment? WWF believe not (Hatton, 2000)
The National Parks and Access to the Countryside Act 1949	Introduced the concept of designation of sites of nature conservation importance	• Designation of National Nature Reserves (NNRs) • Introduced designation of SSSIs (see below) • Conferred powers on local authorities to create Local Nature reserves (LNRs)	• All NNRs are also SSSIs • In March 1998 there were 191 NNRs in England covering 73 374 hectares • In March 1998 598 LNRs in England covering 29 032 hectares had been notified to English Nature (English Nature, 1998a)

Policy/ legislation	Objective	Key provisions	Current position
Wildlife and Countryside Act 1981 (as subsequently amended)	Introduced to address the problem of species protection and habitat loss. The main piece of UK wildlife legislation. Implements provisions of the Birds and Habitats Directives in the UK	• Notification of Sites of Special Scientific Interest (SSSIs) SSSIs given certain protection against damaging operations • Protection of species outside SSSIs • Establishes Areas of Special Protection for Birds (AOSPs)	• Relies heavily on the 'voluntary principle' ie goodwill of landowners not to damage important sites. • At 31 March 1998 there were 3 987 SSSIs in England covering 967 365 hectares (English Nature, 1998). There are currently 1 433 in Scotland covering a total area of 914 029 hectares, 11.6% of the land (Scottish Natural Heritage website at http://www.snh.org.uk – accessed on 16 November 1999) • The Countryside and Rights of Way Bill (Part III) currently before Parliament aims to give SSSIs better protection and management (see the DETR and HMSO websites at http://www.detr.gov.uk and http://www.parliament.the-stationery-office.co.uk). The Scottish Office has recently undertaken a review of the SSSI system – the discussion document *People and Nature: A new approach to SSSI designation in Scotland* was published in 1998 (available via the Scottish Natural Heritage website as above)
Nature Conservation & Amenity Lands (NI) Order (N Ireland) 1985	This implements similar provisions to the Wildlife and Countryside Act 1981 in Northern Ireland	• Sets out the duties of public bodies • Declaration of NNRs, Marine Nature Reserves (MNRs), Areas of Special Scientific Interest (ASSIs) and District Council Nature Reserves	

Policy/ legislation	Objective	Key provisions	Current position
• Planning Policy Guidance (PPG) 9: Nature Conservation issued October 1994 • Scottish Office National Planning Policy Guidance (NNPG) 14 and Circular 6/1995 Nature Conservation issued in 1995. • Planning Guidance (Wales) Technical Advice Note 5: Nature Conservation and Planning (TAN 5), issued in 1996. • Northern Ireland Planning Policy Statement (PPS) 2: Planning and Nature Conservation, issued in 1997	Gives 'guidance on how the Government's policies for the conservation of our natural heritage are to be reflected in land use planning. It embodies the Government's commitment to sustainable development and to conserving the diversity of our wildlife'	• The Government's objectives are 'to ensure that its policies contribute to the conservation of the abundance and diversity of British wildlife and its habitats, or minimise the adverse effects on wildlife where conflict of interest is unavoidable, and to meet its international responsibilities and obligations for nature conservation' (para. 2)	• Acknowledges the importance of undesignated areas for nature conservation • Advises on the treatment of nature conservation issues in development plans • States development control criteria particularly for SSSIs • Contributes to the implementation of the Habitats Directive
	Non-Statutory Nature Reserves	• Designation of non-statutory nature reserves	• Established and managed by a variety of public and private bodies eg County Wildlife Trusts, RSPB
	Sites of Importance for Nature Conservation (SINCs), Sites of Nature Conservation Importance (SNCIs), or their equivalent	• Designation of Sites (and sometimes corridors) of Importance for Nature Conservation	• Usually adopted by local authorities for planning purposes. The name and status of this type of site varies considerably

Appendix 2 – National HAP habitats where road developments are likely to be a factor causing loss or decline

National HAPs	Current factors causing loss or decline
Reedbeds	• Small total area of habitat and critically small population sizes of several key species dependent on the habitat • Pollution of freshwater supplies to the reedbed
Saline lagoons	• Pollution from direct inputs to the lagoon or from the water supply to the lagoon
Chalk rivers	• Direct destruction as a result of development pressure
Fens	Small total area of habitat and critically small population sizes of several key species dependent on the habitat
Ancient and/or species-rich hedgerows	• Removal for development purposes
Lowland heath	• Fragmentation and disturbance from developments such as housing and **road constructions**
Coastal and floodplain grazing marsh	• Localised effects from industrialisation and urbanisation
Purple moor grass and rush pastures	• Fragmentation and disturbance from developments such as housing and **road constructions**
Upland oakwood	• Development pressures such as new roads and quarrying
Native pine woodlands	• Fragmentation and isolation of individual woods with consequent loss of wildlife interest and possibly loss of genetic variation
Mesotrophic Lakes	• Pollution
Aquifer fed naturally fluctuating water bodies	• **Road drainage** may result in over-enrichment of the lake water with plant nutrients (eutrophication), leading to algal blooms and loss of biodiversity
Eutrophic standing water	• Pollutants from diffuse sources
Lowland meadows	• The factors currently affecting lowland meadows (which include atmospheric pollution) reduce the quality and decrease the quantity of the habitat, and its fragmentation brings increased risk of species extinctions in the small remnant areas
Upland hay meadows	• The factors currently affecting upland hay meadows (which include atmospheric pollution) reduce the quality and decrease the quantity of the habitat, and its fragmentation brings increased risk of species extinctions in the small remnant areas
Lowland dry acid grassland	• The factors currently affecting acid grassland (which include development activities such as **road building**) reduce the quality and decrease the quantity of the habitat, and its fragmentation brings increased risk of species extinctions in the small remnant areas
Lowland calcareous grassland	• The factors currently affecting calcareous grassland (which include development activities such as **road building**) reduce the quality and decrease the quantity of the habitat, and its fragmentation brings increased risk of species extinctions in the small remnant areas
Lowland wood-pasture and parkland	• Changes to ground-water levels leading to water stress and tree death resulting from activities including **roads** • Isolation and fragmentation of the remaining parklands and wood-pasture sites in the landscape. (Many of the species dependant on old trees are unable to move between these sites due to their poor powers of dispersal and the increasing distances they need to travel) • Pollution derived from **traffic**

National HAPs	Current factors causing loss or decline
Wet woodland	• Clearance and conversion to other land uses, particularly in woods recently established on wetland sites
Lowland beech and yew woodland	• Fragmentation of the habitat as a result of development
Maritime cliff and slopes	• Development being built too close to cliff-tops
Coastal saltmarsh	• Piecemeal smaller scale land claim for industry, port facilities, **transport infrastructure**, and water disposal is still comparatively common. Such developments usually affect the more botanically diverse upper marsh and land ward transition zones
Mudflats	• Land claim for urban and **transport infrastructure** and for industry. Although land claim has slowed considerably in recent years, it has not stopped
Tidal rapids	• Replacement of ferries by bridges and causeways carrying **roads**
Mudflats in deep water	• Construction of **roads**, bridges and barrages may affect the local hydrodynamic and sediment transport regimes of inshore enclosed areas and consequently affect the deep mud substratum
Lowland raised bog	• Built development – linear developments and other land reclamation for built development may affect many areas

(HM Government, 1995b; English Nature, 1998c, 1999c and d)

Bog

Appendix 3 - National SAP species where road developments are likely to be a factor causing loss or decline

National SAPs	Current factors causing loss or decline
Mammals	
Water vole (*Arvicola terrestris*)	• Loss and fragmentation of habitats
Otter (*Lutra lutra*)	• Pollution of watercourses • Incidental mortality, primarily by **road deaths** and drowning in eel traps
Red squirrel (*Sciurus vulgaris*)	• Habitat fragmentation
Barbastelle bat (*Barbastella barbastellus*)	• Threats to this species are poorly understood, but its low population density and slow population growth make it particularly vulnerable to factors such as: • Further loss and fragmentation of ancient deciduous woodland habitat • Loss, destruction and disturbance of roosts or potential roosts in buildings, trees and underground sites
Bechstein's bat (*Myotis bechsteinii*)	• Threats to this species are poorly understood, but its low population density and slow population growth make it particularly vulnerable to factors such as: • Further loss and fragmentation of open ancient deciduous woodland habitat • Loss, destruction and disturbance of roosts or potential roosts (particularly in old trees)
Lesser horseshoe bat (*Rhinolophus hipposideros*)	• Further loss, damage and fragmentation of woodland foraging habitat, old hedgerows and tree lines, and other appropriate habitat
Birds	
Bittern (*Botaurus stellaris*)	• Degradation of habitat through water pollution
Corncrake (*Crex crex*)	• Disturbance may be contributing to the decline in some localities
Nightjar (*Caprimulgus europaeus*)	• The area of heathland in the UK has undergone a dramatic reduction during the course of this century due to agricultural land claim, afforestation and built development. Eg it is estimated that 40% of England's lowland heathland has been lost since the 1950s. Threats continue from housing and infrastructure development
Cirl bunting (*Emberiza cirius*)	• Habitat loss. Built developments, removal of hedges…have resulted in the loss of cirl bunting breeding and wintering sites
Woodlark (*Lullula arborea*)	• An estimated 40% of England's lowland heathland has been lost since the 1950s. This has lead to a loss of feeding and nesting habitats for woodlarks. Whilst losses to afforestation and agriculture have declined, threats from **roads** and housing developments continue
Black grouse (*Tetra tetrix*)	• Fragmentation of black grouse habitat often leads to small populations which are unlikely to persist
Reptiles and amphibians	
Sand lizard (*Lacerta aglis*)	• Loss, deterioration and fragmentation of heathland and dune habitat to a wide variety of competing uses and pressures eg development
Great crested newt (*Triturus cristatus*)	• Loss of suitable breeding ponds caused by various activities including water table reduction, infilling for development, loss and the degradation, loss and fragmentation of terrestrial habitats
Pool frog (*Rana lessonae*)	• Reduction in the number and quality of suitable ponds in close proximity to each other. This can be caused by several factors including atmospheric pollution and lowered water tables
Fish	
Twaite shad (*Alosa fallax*)	• Pollution • Habitat destruction
Invertebrates	
Coleoptera	
Agabus brunneus (a diving beetle)	• Damage to headwater drainage systems, particularly associated with tourist development and **road improvement**
Amara famelica (a ground beetle)	• Loss of heathland
Anisodactylus poeciloides (a ground beetle)	• Loss of coastal saltmarshes to urban, industrial or recreational developments

National SAPs	Current factors causing loss or decline
Heath tiger beetle (*Cicindela sylvatica*)	• Loss of heath
Crypocehalus primarius (*a leaf beetle*)	• Loss of calcareous grassland
Crypocehalus sexpunctatus (*a leaf beetle*)	• Loss of broadleaved woodland
Spangled diving beetle (*Graphoderus zonatus*) (only known to occur naturally in Britain in Woolmer Forest, north Hampshire)	• Pollution by increased run-off from neighbouring **roads**
Helophorus laticollis (*a water beetle*)	• Damage to headwater drainage systems, in particular associated with tourist development and *road improvement*
Lesser silver water beetle (*Hydrochara caraboides*)	• Loss of ponds to urban development
Hydroporus rufifrons (*a diving beetle*)	• Loss of unimproved pasture
Pterostichus kugelanni (*a ground beetle*)	• Loss of habitat (heathland with sandy or gravely soil, but with wet areas present)
Hydroporus cantabricus (*a diving beetle*) (Species Statement (SS))	• Loss of heathland habitats through agricultural improvement, afforestation and urban encroachment
Bidessus minutissimus (*a diving beetle*)	• Impoundment, bank strengthening, canalisation and other forms of river regulation
Harpalus froelichi (*a ground beetle*)	• Loss of ruderal communities on disturbed sands, including field margins
Synaptus filiformis (*a click beetle*)	• River engineering operations including channel straightening • Pollution/nutrient levels in river water and sediments maybe a factor
River shingle beetles (*6 species, 3 in the family Carabidae, 1 in the family Hydrophilidae, and 2 in the family Staphylinidae*)	• The species share a specific and in some instances, more or less exclusive association with exposed riverine sediments, mostly of the shingle type • Land changes and development (eg urban) that impinge on riparian habitats
Anisodactylus nemoravagus (*a ground beetle*) SS	• Loss and fragmentation of heathland
Dune tiger beetle (*Cicindela maritima*) SS	• Coastal development, especially for tourism
Harpalus dimidiatus (*a ground beetle*) SS	• Loss of calcareous grassland through agricultural improvement, road building or spread of urban and other land development
Saproxylic beetles (Grouped SS)	• 10 species all associated with dead wood habitats on veteran trees in old deciduous woodlands and parklands. Some are restricted to single sites and/or host species, while others are more widespread and are found on a range of trees • Loss of degradation of old woodlands and parklands, through changes of landuse such as conversion to arable farmland or urban development
Diptera	
Bombylius minor (*heath bee-fly*)	• Loss and fragmentation of heathland habitat, including verge heath, owing to development and scrub encroachment
Cliorismia (=*Psilocephala*) rustica (*a stiletto fly*)	• The removal of sandy sediment from rivers and river banks for aggregate and the deepening and canalisation of water courses
Hammerschmidtia ferruginea (*a hoverfly*)	• Loss of aspen woodlands to **road** and building development
Rhabdomastix laeta (=*hilaris*) (*a cranefly*)	• The removal of sandy sediment from rivers and river banks for aggregate • Deepening and canalisation of water courses
Hornet robberfly (*Asilus crabroniformis*)	• Loss of unimproved grassland and heath leading to habitat fragmentation
Hymenoptera	
Andrena ferox (*a mining beetle*)	• The loss of open grasslands with areas of sunny bare ground at the margins and in the rides of broadleaved woodlands

National SAPs	Current factors causing loss or decline
Banded mining bee (*Andrena gravida*)	• Loss of open areas of sandy ground for nesting, and flower-rich sandy grasslands for foraging
Andrena lathyri (a mining bee)	• The loss of open sites on tall sward calcareous or mesotrophic grasslands supporting large populations of vetches
Great yellow bumble bee (*Bombus distinguendus*)	• Loss of extensive, herb-rich grasslands
Cerceris quadricincta (a solitary wasp)	• Loss of open areas of sandy ground for nesting, and flower-rich sandy grasslands for foraging
Cerceris quinquefasciata (a solitary wasp)	• Loss of open areas of sandy ground for nesting, and flower-rich sandy grasslands for foraging
Black bog ant (*Formica candida*)	• Pollution and eutrophication of watercourses • Potential genetic isolation, inbreeding and loss of genetic fitness
Narrow-headed ant (*Formica exsecta*)	• The loss of suitable heathland due to destruction and inappropriate management eg through urban development • Habitat fragmentation leading to potential inbreeding and loss of genetic fitness in isolated populations
Black-backed meadow ant (*Formica pratensis*)	• Urban development on the heaths and cliff tops around Bournemouth
Scottish wood ant (*Formica aquilonia*)	• Loss of suitable native pine woodland
Red barbed ant (*Formica rufibarbis*)	• Loss of suitable heathland habitat through urban or industrial development, agricultural improvement and afforestation
Homonotus sanguinolentus (a spider-hunting wasp)	• Loss of southern heathland, especially grass-heath
Osima uncinata (a mason bee)	• Loss of sites with dead pine wood and suitable open glades
Chrysura hirsuta (a cuckoo wasp) (SS)	• Loss of dead wood and suitable open glades in Caledonian pine woods
Dark guest ant (*Anergates atratulus*)	• Loss of suitable heathland through urban or industrial development and unsympathetic afforestation
Hairy wood ant (*Formica lugubris*) SS	• Loss of suitable woodland habitat through agricultural clearance, urban or industrial development and unsympathetic afforestation
Shining guest ant (*Formicoxenus nitidulus*) SS	• Loss of suitable scrub and woodland habitat through agricultural clearance, urban or industrial development and unsympathetic afforestation
Southern wood ant (*Formica rufa*) SS	• Loss of suitable scrub and woodland habitat through agricultural clearance, urban or industrial development and unsympathetic afforestation
Lepidoptera	
Speckled footman moth (*Coscina cribaria*)	• The loss of suitable habitat due to a variety of factors including development
Marsh fritillary (*Eurodryas aurinia*)	• Development of habitats Increasing fragmentation and isolation of habitats
Netted carpet moth (*Eustroma reticulatum*)	• **Road widening and maintenance**, and alteration to local hydrology
Silver spotted skipper butterfly (*Hesperia comma*)	• Loss of unimproved calcareous grasslands and fragmentation of remaining fragments
Large blue butterfly (*Maculinea arion*)	• Loss of habitat
Straw belle (*Aspitates gilvaria*)	• Habitat loss due to **road construction**, development, and agricultural improvement of unimproved calcareous grassland
Marsh moth (*Athetis pallustris*)	• Changes in land use including drainage and development
Striped lychnis (*Cucullia lychnitis*)	• Inappropriately timed cutting of the larval food plant (Verbascum nigrum and occasionally other Verbascum and Scrophularia species)
Dingy mocha (*Cyclophora pendularia*)	• Loss of heathland to development, forestry, agricultural improvements and **road construction**
Adonis blue (*Lysandra bellargus*)	• Loss of unimproved calcareous grasslands and fragmentation of remaining habitat
Barberry carpet (*Pareulype berberata*)	• Damage to the food plant (barberry (Berberis vulgaris) by burning, mechanised hedge trimming and hedgerow removal

National SAPs	Current factors causing loss or decline
Silver-studded blue (*Plebejus argus*)	• Loss of heathland to development and agriculture • Fragmentation and isolation of habitat
Four-spotted moth (*Tyta luctuosa*)	• Loss of habitat due to agricultural intensification and development
Toadflax brocade (*Calophasia lunula*) (SS)	• Coastal development, sea defence work and **road-widening projects** threaten remaining habitat (shingle at Dungeness, and less commonly on roadside verges, waste ground and in gardens where the food plant grows in open situations). Larvae food plants chiefly Yellow toadflax (Linaria vulgaris), but also other Linaria spp and on small toadflax (Chaenorhinum minus)
Scarce merveille du jour (*Moma alpium*) (SS)	• Clearance of oak woodlands
Belted beauty (*Lycia zonaria britannica*)	• Land development
Barred tooth-striped (*Trichopteryx polycommata*)	• Loss of downland habitat
Chalk carpet (*Scotopteryx bipunctaria*) SS	• Loss of unimproved calcareous grassland and fragmentation of remaining habitat
Other invertebrates	
White-clawed crayfish (*Austropotamobius pallipes*)	• Habitat modification • Pollution
Freshwater pearl mussel (*Margaritifera margaritifera*)	• Poor water quality including nutrient enrichment • Habitat removal and alteration through development, drainage schemes, flow regulation
Starlet sea anemone (*Nematostella vectensis*)	• Loss and damage to lagoon and other sheltered brackish water habitats caused by pollution, drainage and other activities • Isolation of pools leading to fragmentation of populations
Depressed river mussel (*Pseudanodonta complanata*)	• Threats likely to include water pollution, physical disturbance of river banks and channels
Narrow-mouth whorl snail (*Vertigo angustior*)	• The habitat of this snail is very vulnerable to changes in hydrological conditions, reduced grazing pressure and physical disturbance
Desmoulin's whorl snail (*Vertigo moulinsiana*)	• Destruction of wetlands • Habitat degradation, particularly as a result of changes in hydrology
Fen raft spider (*Dolomedes plantarius*) (Order: Araneae)	• Deterioration in water quality • Loss of suitable wetland habitat
Large marsh grasshopper (*Stethophyma grossum*) (Order: Orthoptera)	• Land use on areas adjacent to occupied sites may also affect this species through pollution and impact on local water tables
Uloborus walckenaerius (*a spider*) (Order: Araneae) (SS)	• Loss of heathland due to development and afforestation
Tadpole shrimp (*Triops cancriformis*) (Order Notostraca)	• Pollution
Flowering plants	
Starfruit (*Damasonium alisma*)	• Loss of habitat through development, drainage and infilling of ponds and wet hollows
Eyebrights (*Euphrasia species endemic to the UK*)	• Loss of habitat, particularly inland heaths in Cornwall
Early gentian (*Gentianella anglica*)	• Loss of suitable habitats on dunes, cliffs and limestone or chalk grassland
Tower mustard (*Arabis glabra*)	• Habitat destruction due to agricultural intensification and building development
Deptford pink (*Dianthus armeria*)	• Conversion of pasture to arable and building land • Destruction of hedgerows
Hawkweeds (*Hieracium sect Alpestria*) presently recorded only in Shetland	• Changes in land use, including house building, quarrying and **road widening**
Marsh clubmoss (*Lycopodiella imundata*)	• Habitat loss through eg building development and improvement of unmade tracks • Atmospheric pollution

Nightjar

National SAPs	Current factors causing loss or decline
Penny royal (*Mentha pulgium*)	• Habitat destruction by agricultural intensification and development
Perennial knawl (*Scleranthus perennis ssp prostratus*)	• Loss of sites due to building developments
Cotswold pennycress (*Thlaspi perfoliatum*)	• Removal of hedges, walls and associated banks
Fungi	
Sandy stilt puffball (*Battarraea phalloides*) at 3 sites only	• Loss of hollow trees which provided its former habitat • **Road-widening or resurfacing** of road at Suffolk site
Tulostoma niveum (*a gasteromycete fungus*) a single colony in Scotland known	• A major road improvement scheme could potentially affect peripheral parts of the population
Threatened 'tooth' (or *stipitate hydnoid*) fungi – 14 species	• Historic losses of native pine wood and wood pasture, and perhaps also recent losses of these habitats to agriculture and building development is likely to have reduced the UK population of these species
Moss	
Cornish path-moss (*Ditrichum cornubicum*) known at one site in Cornwall	• Loss of habitat through re-surfacing and disturbance by **vehicles**
Slender green feather-moss (*Hamatocaulis vernicosus*)	• Lowland heath degradation due to lowering of the water table, water pollution
Triangular pygmy-moss (*Acaulon triquetrum*) recently only seen at one site in East Sussex and two sites in Dorset	• Factors responsible for the decline of this species may include tourist and other building developments
Multi-fruited river moss (*Crypaea lamyana*)	• River-bank engineering work including channel straightening/re-profiling, removal of river bank trees and boulders
Blunt-leaved bristle-moss (*Orthotrichum obtusifolium*)	• The general loss of wayside trees through **road improvement**, and parkland trees through senescence, may also have reduced the amount of available habitat. (This moss is an epiphyte on the trunks of trees with nutrient-rich bark growing in open situations. It grows on elm, sycamore, ash and elder.)
Round-leaved feather moss (*Rhynchostegium rotundifolium*) Recently only recorded from two sites in Sussex and Gloucestershire	• **Major highway improvements** alongside the Gloucestershire site
English rock-bristle (*Seligeria calycina=paucifolia*) SS	• Loss of habitat to building developments, **roads**, or arable conversion
Liverworts	
Norfolk flapwort (*Lophozia rutheana*) recently only recorded from one site in Norfolk	• **Road widening** at the Norfolk site
Petalwort (*Petalophyllum ralfsii*)	• Loss of habitat due to development

(HM Government, 1995b; English Nature, 1998b & c, 1999a, b, c & d)

Notes:
1. Each of the species has a SAP unless SS stated, in which case the species has a Species Statement
2. There may be further additions to this list of species of conservation concern as changes in status become apparent and greater knowledge of species requirements develops

Appendix 4 – Cumulative Effects Assessment References

Cumulative Effects Assessment (CEA) references

Key references – highly recommended

- US Council on Environmental Quality (CEQ) (1997) *Considering Cumulative Effects Under the National Environmental Policy Act,* CEQ. Available on the CEQ website at http://ceq.eh.doe.gov/nepa/nepanet.htm.
- Canadian Environmental Assessment Agency (CEAA) (1994) A *Reference Guide for the Canadian Environmental Assessment Act: Addressing Cumulative Environmental Effects,* CEAA, Canada. Available on the CEAA website at http://www.ceaa.gc.ca.
- Canadian Environmental Assessment Agency (1996b) *Cumulative Environmental Effects Cross-Referenced Annotated Bibliography,* CEAA, Canada. Available on the CEAA website at http://www.ceaa.gc.ca.
- CEAA (1999) *Cumulative Effects Assessment Practitioners Guide,* CEAA, Canada. Available on the CEAA website at http://www.ceaa.gc.ca.

Other selected references

- Barnes, JL & Davey, LH (1999) *A Practical Approach to Integrated Cumulative Environmental Effects Assessment to Meet the Requirements of the Canadian Environmental Assessment Act,* prepared for a workshop on Cumulative Effects Assessment, 19th Annual Meeting of IAIA, Glasgow, Scotland, 16-17 June 1999.
- Burris, R.K. & Canter, L.W. (1997) Cumulative impacts are not properly addressed in environmental assessments, *Environmental Impact Assessment Review* Vol. 17 No 1 pp15-18
- Canter, L.W. & Kamath, J. (1995) Questionnaire checklists for cumulative impacts, *Environmental Impact Assessment Review* Vol. 15 No 4 pp311-340
- Clark, R. (1994) Cumulative Effects Assessment: A Tool for Sustainable Development, *Impact Assessment* Vol. 12 No 3 pp319-331
- Cooper, T.A. & Canter, L.W. (1997) Substantive Issues in Cumulative Impact Assessment: A State-of-Practice Survey, *Impact Assessment* Vol. 15 No1 pp15-31
- Damman, D.C, Cressman, D.R. & Sadar, M.H. (1995) Cumulative Effects Assessment: The Development of Practical Frameworks, *Impact Assessment* Vol. 13 No 4 pp433-454
- McCold, L. & Holman, J. (1995) Cumulative impacts in environmental assessments: how well are they considered? *The Environmental Professional* Vol. 17 No 1 pp2-8
- McCold, L. & Saulsbury, J.W. (1996) Including past and present impacts in cumulative impact assessments, *Environmental Management* Vol. 20 No 5 pp767-776
- Manchester University EIA Centre (1998) *EIA Newsletter 14: Cumulative Impacts and EIA.* Available on the EIA Centre website at http://www.art.man.ac.uk/eia/eiac.htm.
- Smit, B. & Spaling, H. (1995) Methods for cumulative effects assessment, *Environmental Impact Assessment Review* Vol. 15 No 1 pp81-106
- Spaling, H. (1994) Cumulative effects assessment: concepts and principles, *Impact Assessment* Vol. 12 No 3 pp231-252.

Appendix 5 – Evaluation Matrix for Determining Impact Significance

Evaluation matrices (such as the one below (adapted from Warwickshire County Council, 1994, 1996)) can be useful as an aid for determining impact significance. However, the impact significance classifications set out in the matrix need to be applied flexibly. For example, use of the matrix below suggests that it is only possible to have a 'severe' impact on a receptor that is of County-level importance or above. Whereas, in reality, there may be circumstances in which the magnitude of an impact is such that it would constitute a 'severe' impact even on a receptor of 'lesser' importance.

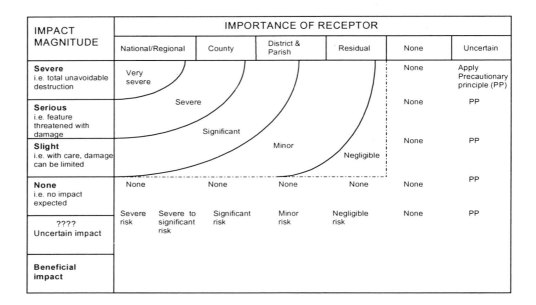

Feedback Form

Please photocopy and return to:

Helen Byron
Environmental Assessment Project Officer
RSPB
The Lodge, Sandy
Bedfordshire SG19 2DL

Tel: 01767 680551
Fax: 01767 683640

Comments on: Biodiversity Impact - Biodiversity and Environmental Impact Assessment: A Good Practice Guide for Road Schemes